Reconciling food law to competitiveness

Reconciling food law to competitiveness

Report on the regulatory environment of the European food and dairy sector

Bernd van der Meulen

with the co-operation of

Isabel Cachapa Rodrigues, Maria Litjens, Adeline Merey,
Giogio Schiavon, Harry Bremmers, Krijn Poppe and Jo Wijnands

Wageningen Academic
P u b l i s h e r s

Also available in the 'European Institute for Food Law series':

European Food Law Handbook
Bernd van der Meulen and Menno van der Velde
ISBN 978-90-8686-082-1
www.WageningenAcademic.com/foodlaw

Fed up with the right to food?
The Netherlands' policies and practices regarding the human right to adequate food
edited by: Otto Hospes and Bernd van der Meulen
ISBN 978-90-8686-107-1
www.WageningenAcademic.com/righttofood

ISBN 978-90-8686-098-2

First published, 2009

© Wageningen Academic Publishers
The Netherlands, 2009

The individual contributions in this publication and any liabilities arising from them remain the responsibility of the authors.

The publisher is not responsible for possible damages, which could be a result of content derived from this publication.

Table of contents

Foreword

Food law matters. The food industry is the most important manufacturing sector in the European Union. Some 17.3 million businesses employ almost 32.5 million people generating a value added of over 650 billion Euros.

The food sector is a heavily regulated sector. In fact it is the third most regulated sector in the EU after automobiles and chemistry. A robust regulatory environment is necessary to ensure food safety for the 500 million consumers in the EU. Previous research (Wijnands *et al.*, 2007) has shown that the competitive position of the EU food industry is declining. To better understand this worrisome development, DG Industry and Enterprises commissioned subsequent research focussing on a specific subsector and giving special attention to the effects of food legislation on the sector. The food sector in general is different from other major manufacturing sectors in the EU in that it consists for 99% (and 50% of turnover) of small and medium-sized enterprises. Can these enterprises carry the weight of such a regulatory burden, the complexity of the regulation, its lack of visibility and its constant changing? Does the regulation in this sector actually achieve what should be the ultimate goal in EU law, which is the free movement of goods in a Single Market, for the benefit of consumers as well as the industry, and the economy in general?

The legal part of the project has been undertaken by Professor Van der Meulen and his team. You are now holding in your hands the results they have come up with, a both loving and critical account of European food law as it stands today and of possibilities to increase harmony between the interests of consumer protection on the one hand and food industry's competitiveness on the other hand. The study applies a combination of legal and empirical methods not usually seen in legal literature.

Shortcomings in EU food law are demonstrated and options for improvement identified. A surprising finding is that many of the problems businesses experience from the regulatory environment are not caused by the law as such but by administrative practices falling short of complying with obligations on authorities such as the obligation to respect deadlines and to provide the reasoning underlying decisions.

Regulatory burdens can be eased by improving administrative practices and by living up to the ambition of 'better regulation'. Private schemes may provide the food sector with opportunities to help itself cope with its regulatory environment, but then particular attention should be given to possible antitrust issues. The sector also needs more clarity as to the respective competence of EU institutions, the national authorities and possible private schemes.

The analyses in this book will deepen the reader's understanding of EU food law and the way it works in practice, the recommendations are fit to help to improve it.

The opinions expressed are those of the authors, not – necessarily – of the European Commission, but they deserve to be considered in future development of European food law.

Nicole Coutrelis

Lawyer at the Paris Bar
President of the European Food Law Association (www.efla-aeda.org)

About the authors

Harry Bremmers
Wageningen University – Business Administration Group, P.O. Box 8130, 6700 EW Wageningen, the Netherlands; Harry.Bremmers@wur.nl

Isabel Cachapa Rodrigues
Wageningen University – Law & Governance Group, P.O. Box 8130, 6700 EW Wageningen, the Netherlands; present adress: Regulatory and Public Affairs Officer at Yakult Europe, Schutsluisweg 1, 1332 EN Almere, the Netherlands; iccrodrigues@gmail.com

Maria Litjens
Wageningen University – Law & Governance Group, P.O. Box 8130, 6700 EW Wageningen, the Netherlands; Maria.Litjens@wur.nl

Adeline Merey
Wageningen University – Law & Governance Group, P.O. Box 8130, 6700 EW Wageningen, the Netherlands

Krijn Poppe
LEI – Wageningen UR, P.O. Box 29703, 2502 LS, The Hague, the Netherlands; Krijn.Poppe@wur.nl

Giogio Schiavon
Wageningen University – Law & Governance Group, P.O. Box 8130, 6700 EW Wageningen, the Netherlands; present adress: Ecor-NaturaSì SpA, Purchase Department, Via Tintoretto 16, Ponte San Nicolò (PD) 35020, Italy

Bernd van der Meulen
Wageningen University – Law & Governance Group, P.O. Box 8130, 6700 EW Wageningen, the Netherlands; Bernd.vanderMeulen@wur.nl

Jo Wijnands
LEI – Wageningen UR, P.O. Box 29703, 2502 LS, The Hague, the Netherlands; Jo.Wijnands@wur.nl

Acknowledgement

Many thanks to Michael Fogden for suggestions on content and language.

Key findings

From a food sector competitiveness point of view several major shortcomings in EU food legislation present themselves. Most of these shortcomings can be resolved by improved compliance with the EC Treaty and the general principles of food law.

- Food legislation has been designed to pursue a limited number of objectives. Among these the objective envisaged in Articles 2, 3 and 157 EC Treaty, to ensure conditions for competitiveness, is missing.
- Food legislation has become too complicated to reach its target audience. Food inspectors and private schemes have the potential to bridge the gap.
- The legislator should take responsibility for simplification and codification of food law.
- Premarket approval schemes impede innovation. Compliance with the principles of risk analysis and precaution, requires that such schemes are imposed only if scientific reasons exist to suspect that a category of food may pose a health risk.
- Practices in the application of premarket approval schemes contribute little to maturing the system.
- Zero-tolerance norms should both from the scientific and competitiveness points of view be applied only as provisional measures and be replaced by more specific levels on the basis of further risk assessment.
- Technically the hygiene package holds sufficient flexibility to accommodate traditional small scale production. Member States' authorities and food businesses are insufficiently aware of the available possibilities.
- The proposed overhaul of labelling legislation requires substantive reorientation by food businesses. Such burden seems justifiable only if the project goes substantially further to solving problems and simplifying legislation than is currently envisaged.
- Good administrative practices to enhance food business competitiveness consist of: compliance assistance, deadline discipline, transparency, compliance with legislation addressing administrative authorities and improved compliance with the duty to give reasons (Article 253 EC Treaty; Article 41(2) EU Charter of Fundamental Rights).

1. Introduction

1.1 Background

In 2005 the European Commission (DG ENTR) issued a study on the competitiveness of the food industry in the European Union.[1] This study is hereafter referred to as 'the first food competitiveness study'. The results of this study have been reported in Wijnands *et al.* (2007). In 2007 the Commission issued a subsequent study, hereafter referred to as 'the second food competitiveness study' with a view to exploring more deeply the results of the first study in the context of a specific sub-sector (dairy).[2] The project resulted in four reports. The present report presents the legal and regulatory findings of this second study. The reports on competitiveness of the EU dairy industry[3] and on administrative burdens[4] are published at www.LEI.nl. The final report has been published by the European Commission in 2009.[5]

1.2 Problem statement

On the basis of the specifications for tender, the following problem statement has been formulated regarding the legal and regulatory part of the project.

Identify food legislation:
- that impedes the placing of a food product on the market with the ensuing consequences for competitiveness;
- raises unjustifiable or unnecessary costs to economic operators, which lead *ipso facto* to a price increase of the end food product;
- prolongs delays prior to the placing on the market of end food product causing detrimental effects on competition of the said industry, whereas this legislation does not bring substantial benefit for public health, consumer protection or environment;
- in the Food Industry sub-sector aimed at improving public health, consumer protection or environment that obviously would not have been adopted or would have been worded differently if genuine impact assessment had been carried out prior to their adoption at Community level.

---

[1] General Invitation to Tender No ENTR/05/075 'Competiveness Analysis of the European Food Industry'.
[2] Call for Tenders No ENTR/2007/020 'Competitiveness Analysis of a Specific Subsector within the European Food Industry, Impact Assessment of the Food Legislation and of International Changes in the Agrofood Market'.
[3] Tacken *et al.* 2009 (in press).
[4] Bremmers *et al.* 2008.
[5] Poppe *et al.* 2009.

In addition, the European Commission is considering developing an EU marking scheme. An inventory of legal options in this regard is also a subject of this project.

1.3 Aim of report

On the basis of the problem statement, the aim of this report is to indicate opportunities for the EU legislator and executive to remove avoidable obstacles for the food industry as a means to reduce regulatory burdens and/or enhance competitiveness.

1.4 Research framework

It is less common in legal research to elaborate on the applied methodologies and the strategies to gather the relevant data. As this report is written in the context of a multidisciplinary research, we believe it to be useful to give a short explanation in this section. The section is summarised in Diagram 1.

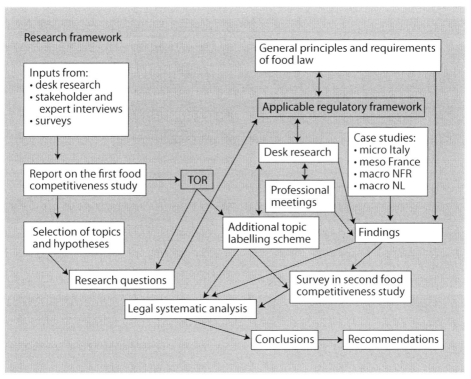

Diagram 1. Research framework.

1.4.1 Methodology

This research has been conducted applying legal systematic research methodologies. These methodologies take a set of assumptions as starting point. A factual assumption is that the different elements of the legal system are connected by the system and mutually influence each other and the system as a whole. Normative assumptions are that the system should respect and apply its own norms and principles and that legislation should be such that it can exercise its intended influence in society. A condition for the fulfilment of the latter assumption is that the relevant elements of the legal system can directly or indirectly be grasped by the addressees.

The unit of analysis of the current study is the system level of EU food law subdivided in subunits of analysis: quality of the system, novel foods premarket approval schemes, food safety targets, food hygiene controls, food labelling and private standards.

1.4.2 Data for desk research

The data for this research have been obtained by desk research supplemented with empirical data. The sampling of empirical data and legislation to be studied was based on the outcomes of the first food competitiveness study. These outcomes have been taken as hypotheses for further testing.

In the first food competitiveness study the following critical issues were identified regarding the impact of food law on the competitiveness of the food sector in the EU:
- quality concerns regarding the food regulatory system as such, like complexity, lack of accessibility and lack of legal certainty due to continuous changes;
- premarket approval requirements;
- density of hygiene controls;
- labelling.

During the preparation of this report, an additional area of concern presented itself: food safety targets. One in-depth interview was dedicated to this issue.

Parallel to this second food competitiveness study, the first author of this report has been involved in the preparation of the 'European Food Law Handbook'[6]. This work added to the identification and selection of relevant documents for desk research. For premarket approval schemes and hygiene controls, the emphasis of the desk research is on the current situation because the system as such was found in the first food competitiveness study to cause concern to food businesses.

---

[6] Nr. 2 in the European Institute for Food Law series: See http://www.wageningenacademic.com/foodlaw (Van der Meulen and Van der Velde, 2008).

Proposals for new legislation have only been taken into account to test whether they provide solutions not present in current legislation. For labelling, the emphasis is on a proposal for restructuring current food labelling legislation because it has been presented as a contribution to competitiveness and is thus of major concern in the current context.

1.4.3 Empirical data

Empirical data have been obtained from the surveys and interviews (63 semi-structured and open interviews) conducted in the context of the first food competitiveness study. After this study three subsequent case studies have been undertaken in the form of MSc research projects conducted at Wageningen University. A fourth case study forms part of a PhD-project at Wageningen University. These case studies yielded further empirical data on the general issues, on novel foods, on food hygiene legislation and controls and on private standards.

The case studies have been spread geographically and by the scale of business. One case study focused on micro-dairy-enterprises producing cheese in a disadvantaged area in Italy, one focussed on regular dairy enterprises producing cheese in France, one focussed on the innovative industry engaged in novel foods and the last addressed the entire dairy chain[7] in the Netherlands.

Giorgio Schiavon conducted a field study in the Italian Alps on the application of the EU hygiene package by traditional micro enterprises. This study will be referred to as 'the Alpeggio case study'. The subjects of this study are micro-enterprises with a limited capacity of production, oriented at local markets and with strong elements of tradition, such as the use of traditional equipment and the custom of *alpeggio*[8] during the good season (itinerant dairy farming). In this case study 34 stakeholders have been interviewed (semi-structured).

Adeline Merey conducted a field study in Champagne, Bourgogne and Normandy (France) on the application of the EU hygiene package by food businesses active in the production of Chaource and Camembert cheese. This study will be referred to as 'the French cheese case study'. In this study 26 persons were interviewed (semi-structured).

Isabel Cachapa Rodrigues addressed the experience gathered in the first ten years of the Novel Foods Regulation. In this study eleven persons have been interviewed (semi-structured). This study will be referred to as 'the NFR case study'.

---

[7] Market structure comprising chains and networks.
[8] There is no equivalent word in English. The definition given by the Lombardia *Health bureau*'s decree n. 6397 is the following: 'the mountain pastures, the shepherd's lodgement, the premises intended for animal husbandry and dairy activities seen as a whole and used for dairy farming during the summer'.

Maria Litjens has mapped the use and role of private standards in the dairy chain in the Netherlands. In this study seven semi-structured interviews have been conducted. This study will be referred to as the 'Dairy Private Regulation case study'.

The majority of interviewees are from food businesses; two are from business organisations, two from DG Sanco, one from a national ministry, two from local authorities and four from national enforcement bodies. These data have been supplemented by results from the survey in the second food competitiveness study. At the time of writing only a very limited number of results had become available (a little over 30). For most questions little more than twenty observations were available.

Additional opinions of experts and stakeholders have been gathered at the European Integrated Project Novel-Q in Spain (Burgos and Lleida) on 23 and 24 October 2007; a workshop organised by the European Food Law Association (EFLA[9]) in Brussels on 11 February 2008, the CIES[10] food safety conference on 14 and 15 February 2008 in Amsterdam, a meeting of the UEAPME[11] working group on foodstuffs on 28 February 2008 in Brussels, the Lebensmittelrechtstage, Wiessbaden Germany on 6 and 7 March 2008, a workshop on the new labelling proposal organised by the Dutch Food Law Association (NVLR[12]) on 12 March 2008, the MLK[13] '"Gesunde" Lebensmittel-Kennzeichnung. Was ist das und gibt es sie überhaupt?' Münster Germany on 31 March 2008, at the seminar 'Regulating Food Safety and Environmental Protection: Legal Challenges', at the University of Copenhagen on 19 and 20 May 2008 and at a meeting with the European Dairy Association at the premises of the European Commission on 7 July 2007.

1.4.4 Data processing

The gathered data have been processed in a qualitative way using as a normative yardstick, the assumption that EU food law should strike a fair balance between food safety and food businesses' interests. At some places in the text findings from the survey in the second food competitiveness study have been presented in numbers and graphs. We do not at this point consider these numbers and graphs to be statistically meaningful, they are presented in support of and illustration of the qualitative findings from the other data sources.

---

[9] See: http://www.efla-aeda.org/.
[10] Comité International d'Entreprises à Succursales (CIES – International Committee of Food Retail Chains) http://www.ciesnet.com/.
[11] UEAPME: Union Européenne de l´Artisanat et des Petites et Moyennes Entreprises, www.ueapme.com.
[12] NVLR: Nederlandse Vereniging voor Levensmiddelenrecht, www.nvlr.nl.
[13] MLK: Münster'sches Lebensmittelrechts-Kolloquium.

1.5 Overview

This report analyses the place of competitiveness in the structure of EU food law in section 2; it addresses the accessibility of food law for its addressees in section 3 including the role of private regulation in annex 1. In section 4 and annex 2 premarket approval schemes are discussed on the basis of an analysis of the novel foods regulation. Section 5 deals with food safety targets. Section 6 focusses on process requirements in general and hygiene law and controls in particular. Its flexibility and impact on the smallest enterprises is studied. Section 7 discusses food labelling legislation, in particular the proposal for a new regulation. Section 8 explores possible supportive labelling schemes. Section 9 takes a look at the role of administrative practices. The discussion concludes in sections 10 and 11 with conclusions and recommendations for the EU legislature and executive.

For the convenience of readers skimming the text, some key findings have been highlighted in boxes. These boxes lack the nuance that is present in the surrounding text.

---

2. Competitiveness and food law

2.1 Introduction

We gave this report the title 'Reconciling food law to competitiveness'. A first question to be addressed in the light of this title is whether any reconciliation is due. Have, in other words, the two (food law on the one hand and competitiveness of the food sector on the other hand) become estranged? To answer this question, in this section we will explore the place of competitiveness in EU law in general and in EU food law in particular.

2.2 Competitiveness in the EC Treaty

Under the heading 'principles' the Treaty of the European Community opens with provisions on the tasks and activities of the European Community. The tasks of the Community are set out in general terms in Article 2 of the EC Treaty. Article 3 lists the activities the Community will undertake to perform its tasks. Further on, the Treaty provisions elaborate on these tasks and activities. In both articles competitiveness holds an important place.

Article 2 EC Treaty:

> '***The Community shall have as its task***, *by establishing a common market and an economic and monetary union and by implementing common policies or activities referred to in Articles 3 and 4,* ***to promote throughout the Community*** *a harmonious, balanced and sustainable development of economic activities, a high level of employment and of social protection, equality between men and women, sustainable and non-inflationary growth,* ***a high degree of competitiveness*** *and convergence of economic performance, a high level of protection and improvement of the quality of the environment, the raising of the standard of living and quality of life, and economic and social cohesion and solidarity among Member States.*'

Article 3(1) EC Treaty:

> '*For the purposes set out in Article 2, the activities of the Community shall include, as provided in this Treaty and in accordance with the timetable set out therein:*
> *(...)*
> *(m) the strengthening of the competitiveness of Community industry;*
> *(...)*'

These general provisions on competitiveness have been fleshed out in Article 157(1) of the EC Treaty.[14] Competitiveness is set as an aim for both the national and Community legislatures:

---

[14] Title XVI Industry.

'1. The Community and the Member States shall ensure that the conditions necessary for the competitiveness of the Community's industry exist.
For that purpose, in accordance with a system of open and competitive markets, their action shall be aimed at:
– speeding up the adjustment of industry to structural changes,
– encouraging an environment favourable to initiative and to the development of undertakings throughout the Community, particularly small and medium-sized undertakings,
– encouraging an environment favourable to cooperation between undertakings,
– fostering better exploitation of the industrial potential of policies of innovation, research and technological development.'

Competitiveness is a key issue of the EU law.

2.3 Objectives of food law

Regulation 178/2002 of the European Parliament and of the Council of 28 January 2002 laying down the general principles and requirements of food law, establishing the European Food Safety Authority and laying down procedures in matters of food safety (the so-called General Food Law[15]) gives the general concepts and principles that serve as the foundation of contemporary EU food law. In this sense the General Food Law is something akin to a 'constitution' of food law in the European Union. The first provision in the section on general principles of food law is Article 5 setting the General objectives of food law. The first two paragraphs of this Article read as follows:

'1. Food law shall pursue one or more of the general objectives of a high level of protection of human life and health and the protection of consumers' interests, including fair practices in food trade, taking account of, where appropriate, the protection of animal health and welfare, plant health and the environment.
2. Food law shall aim to achieve the free movement in the Community of food and feed manufactured or marketed according to the general principles and requirements in this Chapter.'

The concept of 'food law' is defined in a very broad sense. Article 3(1) Regulation 178/2002 gives the definition: *'the laws, regulations and administrative provisions governing food in general, and food safety in particular, whether at Community or national level; it covers any stage of production, processing and distribution of food, and also of feed produced for, or fed to, food producing animals.'* Apparently this concept encompasses *all* legislation (and related case law) relevant to food. To all this legislation, Article 5 GFL applies.

---

[15] Here after also abbreviated as: GFL, see abbreviations.

Article 5 GFL indicates 'objectives' to be pursued, [other interests] to be taken into account and 'free movement of food' to be 'achieved'. This 'pursuing', 'taking into account' and 'achieving' applies to all law within the scope of the definition of food law.

How are these concepts to be understood? From the sequence of the sections in Article 5 GFL, the logic presents itself that the objectives are conditional to the aim. In other words, free movement is to be achieved only for foods that are in compliance with the legislation pursuing protection of health or (other) consumers' interests. European and national legislation on food must *always* target one of these two types of consumer protection, such that this results in the free movement of the products.

What does the element 'taking into account' mean in this context? Article 5 fixes the objectives. This has consequences for interests not mentioned as objectives. These may not be the aim of food law. In most cases objectives can be pursued in different ways. In choosing from the alternatives the other interests may – where appropriate – be taken into account. That is to say in choosing between alternatives to serve consumer interests, preference should be given to options favourable to animal health and welfare, plant health or the environment.

Competitiveness is not among the objectives of EU food law.

Competitiveness and other interests of food businesses and of the food sector are not mentioned; not as objectives to be pursued nor as interests to be taken into account.

2.4 Industry interests

The absence in Article 5 of any mention of the interests of food businesses or of the sector as such is striking.[16] While food legislation in the vast majority of cases addresses food business operators,[17] it is outside the legal scope of this legislation to serve their interests, *or even to take their interests into account.*

---

[16] Given the emphasis the EU places on respect for human rights, the almost complete absence of reference to human rights in EU food law, is maybe even more remarkable. Human rights are relevant for the point at issue. Article 16 of the Charter of Fundamental Rights of the European Union recognises the freedom to conduct a business in accordance with Community law and national laws. While inherent in this text are limitations set by legislation, this legislation should conform to Article 52 of the Charter. The same is true for the numerous limitations to the freedom of expression and information that follow from the provisions on food labelling. Finally it should be borne in mind that adequate (nutritionally sufficient, safe and cultural acceptable) food is in itself a human right recognised in the International Covenant on Economic Social and Cultural Rights (Article 11). See on this issue FAO 2005.

[17] As defined in Article 3(6) Regulation 178/2002.

The objectives of food law as enacted in Article 5 GFL are reflected in DG Sanco's mission statement.[18]

> *'The EU integrated approach to food safety aims to assure a high level of food safety, animal health, animal welfare and plant health within the European Union through coherent farm-to-table measures and adequate monitoring, while ensuring the effective functioning of the internal market.*
> *The implementation of this approach involves the development of legislative and other actions:*
> *To assure effective control systems and evaluate compliance with EU standards in the food safety and quality, animal health, animal welfare, animal nutrition and plant health sectors within the EU and in third countries in relation to their exports to the EU;*
> *To manage international relations with third countries and international organisations concerning food safety, animal health, animal welfare, animal nutrition and plant health;*
> *To manage relations with the European Food Safety Authority (EFSA) and ensure science-based risk management.'*

Somewhat in contrast to this mission statement and its underlying legal imperative, stands the mission statement of the competent authority in the USA, the U.S. Food and Drug Administration (FDA), explicitly including the encouragement of innovation:[19]

> *'The FDA is responsible for protecting the public health by assuring the safety, efficacy, and security of human and veterinary drugs, biological products, medical devices, our nation's food supply, cosmetics, and products that emit radiation. The FDA is also responsible for advancing the public health by helping to speed innovations that make medicines and foods more effective, safer, and more affordable; and helping the public get the accurate, science-based information they need to use medicines and foods to improve their health.'*

The objectives of EU food law do not necessarily contradict the interests of food businesses. In particular assurance of food safety is often seen as a licence to produce and a legal guarantee of a high level of safety will support the reputation of the sector. Nevertheless, their absence in Article 5 GFL and in DG Sanco's mission statement does little to support them.

DG Sanco's mission statement is in line with the legal limits that have been set to the objectives of food law.

Why have the interests of food businesses and the Treaty obligation to promote competitiveness not been included in the objectives of food law in the European Union? A possible explanation can be derived from history. The restructuring

---

[18] Available at: http://ec.europa.eu/food/intro_cn.htm.
[19] Available at: http://www.fda.gov/opacom/morechoices/mission.html.

of food law after 2000, of which Regulation 178/2002 is the first step, was part of the EU reaction to the BSE-crisis and in particular the criticism voiced in the enquiry report[20] to the European Parliament. In this report the Commission's actions have been characterised as follows: '*It has given priority to the management of the market, as opposed to the possible human health risks (...).*'

The explicit aim of the restructuring was to regain consumers' trust and confidence.[21] It seems likely that after being blamed for putting the market before health, the European Commission desired to dissociate itself from the commercial sector. In hindsight, this may have been somewhat overdone.

The consequences may be bigger for food law at Member State level, than at EU level. National food legislation is subordinate to EU legislation, and therefore cannot legally depart from Article 5 GFL. EU legislation is not subordinate to the GFL. Therefore regulations and directives at EU level have the option to overstep the limits set by Article 5 GFL.[22] However already at this point the conclusion may be drawn, that where the legislator made its objectives explicit, it may be expected that the legislator in recreating EU food law actually did apply these objectives and did not pursue interests it is not supposed to pursue, like those of food businesses and competitiveness. In other words, it is not unlikely that provisions disregarding business interests can be found.

The interests of businesses may yet enter the arena through another door: the principle of proportionality.[23] In pursuing its objectives, legislation should not do more damage to other interests than is necessary to achieve the legitimate aim. The European Court of Justice (ECJ) established case law deriving from the principle of proportionality a requirement of scientific proof of the necessity of measures creating trade barriers.

---

[20] Ortega Medina 1997. Criticism on comitology was tough as well: 'By virtue of the opaqueness, complexity and anti-democratic nature of its workings, the existing system of commitology seems to be totally exempt from any supervision, thereby enabling national and/or industrial interests to infiltrate the Community decision-making process. This phenomenon is particularly serious where public health protection is at stake.'

[21] European Commission 2000a.

[22] Indeed disregarding the limits set on the scope of food legislation by Article 5 GFL, DG Sanco in February 2006 published a consultative document in preparation of new food labelling legislation under the title 'Labelling: competitiveness, consumer information and better regulation for the EU'. The proposal that followed from this consultation, Proposal for a Regulation of the European Parliament and of the Council on the provision of food information to consumers COM(2008) 40 final, 30.1.2008, follows suit. The explanatory memorandum states; 'The proposal is in line with the Commission's Better Regulation Policy, the Lisbon Strategy and the EU's Sustainable Development strategy. The emphasis is on simplifying the regulatory process, thus reducing the administrative burden and improving the competitiveness of the European food industry, while ensuring the safety of food, maintaining high level of public health protection and taking global aspects into consideration.'

[23] EC Treaty; Protocol (30) on the application of the principles of subsidiarity and proportionality.

An example of a measure judged disproportionately was the administrative practice of the Danish authorities, consisting of systematic prohibition, because of unknown effects on human health, of marketing of all foodstuffs to which vitamins or minerals have been added, unless such enrichment in nutrients met a need in the Danish population.[24] According to the ECJ, exceptions justified on the grounds set out in Article 30 EC Treaty[25], as they must be interpreted strictly, require a 'detailed assessment, case-by-case' of the effects of a given product. Such evaluation must prove that prohibition of that product is necessary for the protection of the interests referred to in this provision.[26] In this light, the mere absence of a nutritional need in the population is not sufficient to justify an import ban.[27]

Business interests in food law are protected through the principle of proportionality.

A similar approach is found at WTO level in Article XX(b) of the GATT[28] and in the SPS Agreement (Articles 2(2) and 5).[29] This principle has also been adopted in Regulation 178/2002, in Article 6:

'1. In order to achieve the general objective of a high level of protection of human health and life, food law shall be based on risk analysis except where this is not appropriate to the circumstances or the nature of the measure.

2. Risk assessment shall be based on the available scientific evidence and undertaken in an independent, objective and transparent manner.

3. Risk management shall take into account the results of risk assessment, and in particular, the opinions of the Authority referred to in Article 22, other factors legitimate to the matter under consideration and the precautionary principle where the conditions laid

---

[24] ECJ Case 192/01, *Commission v. Denmark*, [2003] ECR I-9693.

[25] Available at: http://eur-lex.europa.eu/LexUriServ/LexUriServ.do?uri=OJ:C:2006:321E:0001:0331: EN:pdf.

[26] ECJ Case 192/01, paras. 46 and 56, see also EFTA Court Case E-3/00, EFTA Surveillance Authority v. Norway, p. 73.

[27] EFTA Court Case E-3/00, para. 28.

[28] Article XX(b) GATT: Subject to the requirement that such measures are not applied in a manner which would constitute a means of arbitrary or unjustifiable discrimination between countries where the same conditions prevail, or a disguised restriction on international trade, nothing in this Agreement shall be construed to prevent the adoption or enforcement by any contracting party of measures:
(b) necessary to protect human, animal or plant life or health.

[29] Article 2(2) SPS Agreement: Members shall ensure that any sanitary or phytosanitary measure is applied only to the extent necessary to protect human, animal or plant life or health, is based on scientific principles and is not maintained without sufficient scientific evidence, except as provided for in paragraph 7 of Article 5.
Article 5(1) SPS Agreement: Members shall ensure that their sanitary or phytosanitary measures are based on an assessment, as appropriate to the circumstances, of the risks to human, animal or plant life or health, taking into account risk assessment techniques developed by the relevant international organizations.

down in Article 7(1) are relevant, in order to achieve the general objectives of food law established in Article 5.'

Implicitly we find in this provision a protection of the interests of food businesses in that it takes as a principle of food law that provisions protecting human health (by setting trade and other barriers to businesses) need to be based on a scientific foundation.

Proportionality requires restrictive measures to be science-based.

In practice, however, we find an ever increasing body of EU food legislation requiring food businesses to provide in advance scientific proof that the product they wish to bring to the market does not have adverse effects on health.

On such an approach, the Commission Communication on the precautionary principle[30] has the following to say:

'Community rules and those of many third countries enshrine the principle of prior approval (positive list) before the placing on the market of certain products, such as drugs, pesticides or food additives. This is one way of applying the precautionary principle, by shifting responsibility for producing scientific evidence. *This applies in particular to substances deemed 'a priori' hazardous or which are potentially hazardous at a certain level of absorption. In this case the legislator, by way of precaution, has clearly* reversed the burden of proof *by requiring that the substances be deemed hazardous until proven otherwise. Hence it is up to the business community to carry out the scientific work needed to evaluate the risk. As long as the human health risk cannot be evaluated with sufficient certainty, the legislator is not legally entitled to authorise use of the substance, unless exceptionally for test purpose* (emphasis added).'

The qualification by the Commission of reversal of proof through premarket approval requirements as an application of the precautionary principle is important. The content of the precautionary principle and conditions for its application have been elaborated in case law. Also the precautionary principle has been recognised as a fundamental principle of food law in Article 7 GFL[31].

The ECJ has ruled that the prior authorisation procedure (applied by Member States) is, in principle, not contrary to Community law, provided that it is 'readily accessible and can be completed within a reasonable time, and, if it leads to a refusal, the decision must be open to challenge before the courts.'[32]

---

[30] European Commission 2000b.
[31] See abbreviations.
[32] ECJ Case C-24/00, para. 26.

Limitations to proportionality consisting of a shift of the burden of scientific proof to businesses can be justified on the basis of the precautionary principle.

2.5 Concluding remarks on competitiveness in EU food law

The objectives of food law do not necessarily clash with the interests of food businesses or competitiveness of the food sector. From the first food competitiveness study, it follows that the level playing field that EU food law provides for the food sector in the EU-27 is much valued. The high level of consumer protection guaranteed by EU food law, may well contribute to the reputation of EU food products both on the EU market and on the world market and thus directly support the market position of the European food sector.

If, however, the EU legislator complies with the instruction to take into consideration only the interests indicated and not other interests like those of food businesses and of competitiveness, the possibility must be taken into account that provisions disregarding business interests are present in current EU food law. The most important protection for business interests is the requirement derived from the principle of proportionality that food law measures must be based on scientific risk assessment, or if this is inconclusive on the precautionary principle.

In the first food competitiveness study some critical issues of food law regarding competitiveness were identified. Those will be further pursued in the report.

3. Accessibility of food law

3.1 Awareness

The first food competitiveness study indicated that food business operators feel rather confident about being aware of food legislation as far as it is relevant for their activities.[33] Experts on the other hand expressed the opinion that the quantity and complexity of EU food legislation has become such that it is next to impossible to acquire adequate understanding without focussed education. In the French cheese case study and in the Alpeggio case study we gave specific attention to interviewees' knowledge of EU food law as applicable to their activities. Generally the picture was confirmed that business operators believe themselves to be well informed. They had however great difficulty in answering control questions. Most interviewees for example seem to believe that food hygiene is specifically about cleanliness.

One third of the interviewees in the Alpeggio case study had not noticed the transition from directives-based national law to the hygiene package consisting of regulations at EU level. Half of the interviewees did not seem to have even rudimentary knowledge of the principles of Hazard Analysis and Critical Control Points (HACCP) nor could they indicate CCPs in their own processes. Nevertheless, the majority of the respondents felt sufficiently informed on the relevant hygiene requisites (by the officials). In the French cheese case study only eight interviewees (out of 26) could give details on hygiene requirements spontaneously.

In the Alpeggio case study, the French cheese case study and the NFR case study, interviewees' understanding of what is required of them turns out to be based in the first place on customers' demands, in the second place on feedback from inspectors and in the third place on information available in the sector often of a 'hear say' level of precision. Sometimes training is available. Such training is mainly about food hygiene.

Informedness varies with business size. Only the biggest companies apply food law as a tool in their business strategies. Such businesses often have an office or at least a staff member responsible for regulatory affairs. In situations where regulatory affairs is part of the responsibilities of a quality assurance manager, the level of knowledge seems adequate for compliance but not sufficient for pro-active strategies. In both situations, account is taken of actual legal texts. This seems less common in smaller businesses.

---

[33] The survey in the second food competitiveness study yielded the same picture: mean 5.5 (N = 30) on a scale from 1 (uninformed) to 7 (well informed). Representatives of the EDA pointed out that they do not at all feel well informed about REACH and that animal by-products legislation is unsuitable for the dairy sector. It seems to work from the presumption that all by-products are solid body parts and not liquids.

The Alpeggio case study provided an example where inspectors apparently not knowing how to apply the EU hygiene package, for lack of implementing measures continued to inspect on the basis of previous legislation. As a consequence, many businesses believed the old legislation still to be in force.

Knowledge of food law in food businesses is limited.

3.2 Consolidation

In the first food competitiveness study a restructuring of the procedure of EU legislation has been advocated such that each legislative trajectory results in an official publication of a consolidated version of the law. Currently official texts usually only indicate changes to be made, leaving the onus to the user and/or (commercial) editors to actually apply these changes to the original text. From such editors the information was received that this task is far from self-evident and often requires them to select interpretations and make choices.

The EU website at EUR-Lex provides a service of consolidated versions, disclaiming however their reliability.[34] The text most used in this report – the Novel Foods Regulation (Regulation 258/97) – may stand as an example[35] that making a consolidated version is not an easy task. In the text two amendments have been incorporated.[36] Each of these explicitly mentions Regulation 258/97 as a law being

---

[34] The very first line of each document reads: 'This document is meant purely as a documentation tool and the institutions do not assume any liability for its contents.'

[35] Another recent example follows from Commission Directive 2008/5/EC of 30 January 2008 concerning the compulsory indication on the labelling of certain foodstuffs of particulars other than those provided for in Directive 2000/13/EC of the European Parliament and of the Council. It entered into force on 20 February 2008. Article 2 of this directive starts by stating: 'Directive 94/54/EC, as amended by the Directives listed in Annex II, Part A, is repealed, (...)'. This cryptic wording makes sense only if it means that the amending directives (96/21 and 2004/77) are repealed as well. Eur-lex, however, still (at 3.9.2008) mentions them as in force.

[36] Regulation 1829/2003 and Regulation 1882/2003.

amended. An amendment hidden in the General Food Law,[37] however, applying another approach, has been overlooked. Article 62 GFL states among other things that every reference in Community legislation to the Scientific Committee on Food shall be replaced by a reference to the European Food Safety Authority and every reference in Community legislation to the Standing Committee on Foodstuffs shall be replaced by a reference to the Standing Committee on the Food Chain and Animal Health (SCFCAH). Nevertheless in the consolidated version we still encounter the Scientific Committee on Food (Article 11 GFL) and the Standing Committee on Foodstuffs (Article 12 GFL). Article 13 GFL has been updated to mention the SCFCAH.

This is not to criticise the very much valued service provided by EUR-Lex but to illustrate that in the system of legislation applied in the EU uncertainty about the actual text of the law in force is inherent.[38] The legislature should not leave it to somebody else to resolve this issue. Like the US legislator,[39] the EU legislator should accept the administrative burden and codify the law in official binding texts.[40]

Consolidated versions of EU legislation are not official and therefore cannot guarantee reliability.

---

[37] The first 3 paragraphs of article 62 GFL read:
1. Every reference in Community legislation to the Scientific Committee on Food, the Scientific Committee on Animal Nutrition, the Scientific Veterinary Committee, the Scientific Committee on Pesticides, the Scientific Committee on Plants and the Scientific Steering Committee shall be replaced by a reference to the European Food Safety Authority.
2. Every reference in Community legislation to the Standing Committee on Foodstuffs, the Standing Committee for Feedingstuffs and the Standing Veterinary Committee shall be replaced by a reference to the Standing Committee on the Food Chain and Animal Health.
Every reference to the Standing Committee on Plant Health in Community legislation based upon and including Directives 76/895/EEC, 86/362/EEC, 86/363/EEC, 90/642/EEC and 91/414/EEC relating to plant protection products and the setting of maximum residue levels shall be replaced by a reference to the Standing Committee on the Food Chain and Animal Health.
3. For the purpose of paragraphs 1 and 2, 'Community legislation' shall mean all Community Regulations, Directives and Decisions.
[38] Naturally this provides business opportunities for some. The Veterinary Imports Legislation website (www.vetimpleg.eu) advertises its product stating among other things: ' ... sources for EU legislation are not particularly user-friendly, are often not up-todate and can be incomplete. The legislation website of the European Commission, Eurlex, contains all the EU legislation but not in a consolidated way.'
[39] All US statutes are codified in the United States Code (http://www.gpoaccess.gov/uscode/index.html) and all regulations in the Code of Federal Regulations (http://www.gpoaccess.gov/cfr/index.html). Unlike EUR-lex, the text of the Code is legal evidence of the law (Curtis *et al.*, 2005, p. 13).
[40] The European Commission sometimes refers to an instruction to the Commission's staff dated 1 April 1987 (COM(87) 868 PV – not available on Eur-Lex) that all legislative texts should be codified after no more than ten amendments, stressing that this is a minimum requirement and that departments should endeavour to codify at even shorter intervals the texts for which they are responsible, to ensure that the Community rules are clear and readily understandable.

3.3 Controls & businesses

The observation made in the first food competitiveness study that businesses feel overburdened by a stack of different layers of public and private food controls, did not repeat itself in the case studies.

The Alpeggio case study, the French cheese case study and the NFR case study show that the understanding of the law is greatly influenced by the feedback from inspectors. The Alpeggio case study showed that to most interviewees 'the law' and 'inspectors' opinions' are the same thing. The Alpeggio case study indicates that enforcement officers' different approaches and understandings had contributed, to a certain extent, to a non-homogeneous (continued) use of traditional methods in the territory covered by the case study.

3.4 Self-regulation

One of the findings of the first food competitiveness study was that private food standards are helpful in complying with public law requirements. To a certain extent this is surprising. As far as HACCP-requirements are concerned, private standards do indeed elaborate on what has to be done in practice to comply with these requirements. For most other public law requirements, however, most private standards do little more than stating that compliance is due.

The Dairy Private Regulation case study set out to shed additional light on the role and impact of private schemes in the Dutch dairy sector. Annex 1 maps private regulation in the dairy chain in the Netherlands. The emerging picture is very complex. Many different schemes exist, most of these are interrelated. At first sight it seems unlikely that such a complex structure is seen as helpful in complying with public law requirements.

Small players in the production chain will usually be related to a small number of suppliers and customers. To the extent that this is true, these small players will only be confronted with a relatively small part of private regulation.

The real help private standards supply in complying with food law, however, is not primarily from the content of the standard as such but from its infrastructure embedded in training, auditing and certification. This is in line with the findings indicated above that feedback from customers and inspectors form important sources of knowledge about food law. Private certification schemes embed and enhance both sources of feedback. The French cheese case study showed that product rejections by customers usually lead to consultation between producer and customer on strategies to prevent future problems, in particular in the Chaource, less so in the Camembert.

Private standards help to ensure compliance with public law requirements.

The responses from the survey in the second food competitiveness study indicate that stakeholders value the different private schemes differently as supports to comply with food legislation (Diagram 2). Certified HACCP systems are seen as most helpful to complying with public law requirements followed by retailer systems like BRC and GlobalGAP, but ISO and SQF less so.

Compliance with government requirements seems to be an important motive for applying food safety and quality systems (N 25, mean 5.32). Even more important is that customers demand it (N 25, mean 6.24).

The French cheese case study gives ambiguous results regarding the effect of private law incentives on public law compliance. On the one hand are the requirements from a processor applying a protected designation, seen as a stronger incentive for hygienic practices than legislation. On the other hand it is noticed that hygiene guides[41] are not being used in everyday practice. They are considered too complicated. Their message has to be simplified for use in the businesses.

In the Chaource area the dairy companies invest more in enhancing producers' quality than in the Camembert area. In the Chaource there is a decreasing production sector making raw materials become more scarce. The Camembert area shows a more competitive production market. Dairies leave it to the producers to find their own solutions if problems occur.

Diagram 2. Helpfulness of private standards for compliance with food legislation (1 = not helpful at all; 7 = very helpful).

	Certified HACCP	ISO	Retailers' systems (BRC, GlobalGAP)	IFS	SQF	Other systems
N Valid	22	18	23	23	17	6
Mean	6.14	3.78	5.59	4.39	2.94	3.17
Std. deviation	1.125	2.625	1.843	2.445	2.015	2.563

---

[41] Even though they are strongly embedded in public law, hygiene guides are private regulation as well.

3.5 Organisation

The French cheese case study showed that being part of a group is very helpful in achieving understanding of legal requirements. When companies own different establishments, people have to meet on a regular basis and discuss issues of common interest. One of the interviewees in this case study put this observation in perspective by pointing to its limits. If the group is too big, communication is lost. Information from headquarters is ignored if it is believed to come from people who are not aware of the 'real' needs.

3.6 Conclusion on accessibility of EU food law

The findings of the first food competitiveness study have been confirmed, that food businesses feel they have a sufficient working knowledge of EU food law. In the first food competitiveness study, experts expressed the opinion that business' confidence in this respect is not justified. The experts' opinion is supported by the results of the case studies.

Personal contact in the context of official controls and private regulation forms a very important source of food law knowledge and understanding for businesses.

4. Premarket approval schemes

4.1 The first food competitiveness study on premarket approval

The issue of premarket approval requirements has been addressed in Wijnands *et al.* (2007). The findings are presented in the following paragraphs.

Most interviewees agreed that the premarket approval procedures for additives, novel foods, GMOs and (health) claims are beyond reach for the vast majority of food businesses in the EU. Legislation reserves this type of innovation to the happy few. But even for them, life is not easy. Each premarket approval requirement has its own procedure. Harmonisation is limited. If businesses choose the wrong procedure, they cannot simply switch, but have to start all over again.

No help from the authorities can be expected in finding the right procedure or applying it successfully. Interviewees worry whether the authorities will meet their deadlines. One of them provided an example of a procedure that took fourteen years to complete. One of the problems perceived by interviewees is uncertainty on the range of premarket approval schemes. Does the application of a certain preparation technique bring the food within the ambit of the Novel Foods Regulation? Does this application of a genetically modified organism in processing bring the food within the ambit of GM legislation? Is this information concerning the product a claim under the Claims regulation? Etc.

Interviewees advise devising a simple procedure to answer preliminary questions. In particular the decision that a certain procedure does not apply (negative clearance) can be most helpful to open up to innovation. DG Competition has developed an informal form, the so-called comfort letter, in which the Commission gives its interpretation of the legal situation. Similar practices would be most welcome in food legislation.

In short: the problems encountered relate to costs, time and legal certainty. Here we will focus on the latter.

Premarket approval schemes exclude the majority of food businesses from the innovation concerned.

4.2 Quantity

In the period 2003-2008 a total of 25 genetically modified foods were approved, none of them dairy. Nineteen novel foods were approved, four of which are dairy products. Some 30 new additives have been approved, most of them with uses unrelated to the dairy sector.

It seems that less than one hundred innovations in the food sector requiring premarket approval have actually been approved under EU food law. A handful of these apply to the dairy sector. This small quantity is in striking contrast to the total number of innovations reported for the dairy sector. The second food competitiveness study identified over 1,400 innovations in the period 2003 – 2008 for the dairy sector alone.[42] It has been shown that the vast majority of these innovations is outside the scope of product innovation, the area most prone to be affected by premarket approval requirements. By themselves these numbers prove nothing, but they certainly do not disprove stakeholders' view that premarket approval schemes pose barriers to innovation.

4.3 Proliferation of positive list requirements

As indicated above, the principle of proportionality as well as Articles 6 GFL[43] and 5 SPS[44] require scientific risk assessment to justify measures that restrict food businesses. However some leeway is granted to the legislator on the basis of precaution to consider certain substances as '*a priori* hazardous' and to reverse the burden of scientific proof regarding their safety.

An early example in EU food law of such reversal of the burden of proof of safety can be found in legislation on food colours. 'EEC Council Directive on the approximation of the rules of the Member States concerning the colouring matters authorized for use in foodstuffs intended for human consumption,'[45] set out to harmonise Member States' legislation by establishing a single list of colouring matters whose use is authorised for colouring foodstuffs and laying down criteria of purity which those colouring matters must satisfy. Thus one of the first lists was created of food ingredients of a certain type that could be used, while those not on the list could not be used. This is a so-called positive list. The law does not say what is forbidden leaving the rest free, but says what is allowed making the rest forbidden. The list is a part of the law (in this case an annex to the directive). To later include a product in the list (or delete a product from the list) the law must be changed by the applicable procedure.[46]

While the details differ greatly, this system of positive lists set by the law is still the core mechanism of premarket approval schemes in EU food law. Ever since, the

---

[42] See: Poppe *et al.* 2009, Chapter 5.
[43] See abbreviations.
[44] Available at: http://www.wto.org/english/docs_e/legal_e/15-sps.pdf
[45] The uniform numbering system of European legislation did not yet exist. The directive was published in: OJ 115, 11.11.1962, p. 2645–2654.
[46] Unless the law itself provides for a specific procedure.

number of types of food products deemed '*a priori* hazardous' and made subject to a positive list system has been expanded more and more.

In this context German scholars observe a shift in food law from what they call the 'principle of abuse' to the 'prohibition principle with reservation of permission'.[47] If the former principle applies, food businesses are free in their actions but will be held responsible if they infringe on the general norm of food safety.[48] In other words the food is considered not to be categorically unsafe. If the latter principle applies, it is forbidden to bring the food to the market unless an express permission has been obtained. The food is considered categorically unsafe (until authorities decide otherwise). These scholars criticise this development. From the point of view of business competitiveness, the shift of the burden of scientific proof from the authorities where it was placed for reasons of proportionality to businesses, may indeed raise some concern.

A priori hazardous foods need to acquire a place on a positive list to be admitted to the market.

Positive lists or similar premarket approval requirements exist in EU food law as well as in Member States' food law. At EU level they currently apply – at least[49] – to food additives[50] (divided into sweeteners,[51] colours[52] and other ('miscellaneous') additives[53]), extraction solvents,[54] some flavourings,[55] infant formulae[56] and some other foods for particular nutritional uses,[57] food supplements,[58] novel foods,[59]

---

[47] Translation by Will and Guenther, 2007, p. 16.

[48] Now Article 14 GFL.

[49] This is not the place to attempt an exhaustive inventory.

[50] Additives Framework Directive (89/107) on the approximation of the laws of the Member States concerning food additives authorised for use in foodstuffs intended for human consumption.

[51] Sweeteners Directive (94/35).

[52] Colours Directive (94/36).

[53] Miscellaneous Additives Directive (95/2).

[54] Council Directive of 13 June 1988 on the approximation of the laws of the Member States on extraction solvents used in the production of foodstuffs and food ingredients (88/344/EEC).

[55] Framework Directive 88/388; Council Decision 88/389; Directive 89/107; Regulation 2232/96; Commission Decision 1999/217 adopting a register of flavouring substances; Regulation 2065/2003 on smoke flavourings;

[56] Directive 91/321 on infant formulae and follow-on formulae.

[57] Directive 89/398; Directive 2001/15 on substances that may be added for specific nutritional purposes in foods for particular nutritional uses.

[58] Directive 2002/46/EC of the European Parliament and of the Council of 10 June 2002 on the approximation of the laws of the Member States relating to food supplements.

[59] Novel Foods Regulation 258/97.

genetically modified foods,[60] novel food contact materials[61] and decontaminants.[62] One could argue that a premarket approval requirement also applies to functional foods. Functional foods can be seen as foods that are being marketed with the claim that they provide a health benefit beyond basic nutrition. While the food as such does not need to be approved for the EU market, the *claim* may only be made in compliance with the Regulation on nutrition and health claims.[63] Also the use of certain other messages must be approved in advance: protected designations of origin, protected geographical indications[64] and traditional specialities.[65] Even voluntary labelling of beef and veal must be pre-approved.[66]

The system of positive lists is expanding.

4.4 Coming legislation

On 28 July 2006, the European Commission submitted a package of four proposals on food improvement agents to the Council and the European Parliament. This package aimed to harmonise some of the procedures for premarket approval, to recast the legislation on additives and to introduce a premarket approval requirement for flavourings[67] and enzymes. A little later a proposal was published to recast the Novel Foods Regulation.[68] In these proposals a separation is made between procedural matters governed by a common procedures regulation and substantive matter governed by what are called sectoral food laws. From each

---

[60] The GM package encompasses; Directive 2001/18 on the deliberate release into the environment of genetically modified organisms; Regulation 1829/2003 on GM food and feed; Regulation 1830/2003 on traceability and labelling of GMOs and the traceability of food and feed products from GMOs; Commission Regulation 65/2004; Commission Regulation 641/2004; Directive 90/219/EEC on the contained use of genetically modified micro-organisms; Regulation 1946/2003 on transboundary movements of genetically modified organisms and Commission Recommendation C (2003) 2624 on guidelines for national strategies and best practices to ensure coexistence provides Member States with policy options to protect conventional agriculture from GM admixture.

[61] Regulation 1935/2004.

[62] Article 3(2) first sentence Regulation 853/2004: Food business operators shall not use any substance other than potable water – or, when Regulation 852/2004 or this Regulation permits its use, clean water – to remove surface contamination from products of animal origin, unless use of the substance has been approved in accordance with the procedure referred to in Article 12(2).

[63] Regulation 1924/2006.

[64] Regulation 510/2006.

[65] Regulation 509/2006.

[66] Regulation 1760/2000.

[67] Currently for flavourings the inverse applies. 'Council Directive of 22 June 1988 on the approximation of the laws of the Member States relating to flavourings for use in foodstuffs and to source materials for their Production' (Directive 88/388/EEC) addresses contaminants. It applies negative lists holding substances and quantities that should not be present in the final food product.

[68] European Commission 2007.

sectoral food law follows a (positive) list of products approved through the common procedure.

On the last day of 2008 (31 December), four regulations appeared in the Official Journal. Regulation 1331/2008 of 16 December 2008 establishing a common authorisation procedure for food additives, food enzymes and food flavourings; Regulation 1332/2008 of 16 December 2008 on food enzymes; Regulation 1333/2008 of 16 December 2008 on food additives, and Regulation 1334/2008 of 16 December 2008 on flavourings and certain food ingredients with flavouring properties for use in and on foods. Their application is set on different dates after 20 January 2010, some depending on implementing measures that must be taken. The Novel Foods proposal is still in procedure.

All measures in this package are regulations. For food additives and flavourings this means that harmonised national law of the Member States will be replaced by uniform EU law. This development is in line with the general trend in EU food law from directives to regulations. Introduction of a positive lists' system for food supplements other than vitamins and minerals is possible, bur currently not very likely.[69]

4.5 Procedures and consequences

The premarket approval schemes indicated above differ considerably in procedure, competences in assessment and decision, criteria and legal consequences. Diagram 3 gives an overview of some of the competences to deal with applications for approval.

Legal consequences come in two categories: generic and exclusive. Generic approvals address the food product: once approved all food businesses are entitled to market the product. Exclusive authorisations address the applicant; after approval the applicant has a right to exclude all others in bringing the product to the market. Other businesses that want to bring the same innovation to the market also have to pass an approval procedure.[70] The exclusive authorisation is the 'bonus' rewarding businesses for their investment in completing the procedure. From a risk management point of view, it makes little sense to demand scientific risk

---

[69] Article 4 (8) Directive 2002/46: 'Not later than 12 July 2007, the Commission shall submit to the European Parliament and the Council a report on the advisability of establishing specific rules, including, where appropriate, positive lists, on categories of nutrients or of substances with a nutritional or physiological effect other than those referred to in paragraph 1, accompanied by any proposals for amendment to this Directive which the Commission deems necessary.' The deadline mentioned in this provision has not been met. The report appeared on 5 December 2008 (COM(2008) 824 final). It concludes that laying down specific rules applicable to substances other than vitamins and minerals for use in food supplements is not justified.

[70] Or in the case of novel foods a less heavy notification.

Diagram 3. Competences to deal with applications for approval.

	Receive application	Risk assessment	Decision		
Additives (including sweeteners and colours)	Commission	EFSA	Commission, Parliament, Council		
Food supplements	Commission	EFSA	Commission & SCFCAH		
Novel foods initial assessment	National competent authority	National risk assessment body	National competent authority		
Novel foods additional assessment		EFSA	Commission & SCFCAH		
GM environmental approval	National competent authority	National risk assessment body	Assessment is negative: National competent authority	Assessment favourable no objections: National competent authority	Assessment favourable objections: Commission & SCFCAH
GM food approval	National competent authority	EFSA	SCFCAH favourable: Commission	SCFCAH not favourable: Council	SCFCAH not favourable, but Council misses 3 month deadline: Commission

assessment for products known to be safe on the basis of the evidence submitted by another business.[71] Indeed it seems incompatible with the precautionary principle to require scientific proof of safety in situations where no scientific uncertainty exists because risk assessment has already been conducted.

Approval requirements for foods known to be safe from previous risk assessment cannot be based on the precautionary principle.

---

[71] This situation occurs in particular with regard to so-called 'exotic' foods. These are foods new to the EU originating from other parts of the world.

In Article 7 GFL the precautionary principle as applied to food law has been enacted as follows:

'*Article 7*

Precautionary principle

1. In specific circumstances where, following an assessment of available information, the possibility of harmful effects on health is identified but scientific uncertainty persists, provisional risk management measures necessary to ensure the high level of health protection chosen in the Community may be adopted, pending further scientific information for a more comprehensive risk assessment.

2. Measures adopted on the basis of paragraph 1 shall be proportionate and no more restrictive of trade than is required to achieve the high level of health protection chosen in the Community, regard being had to technical and economic feasibility and other factors regarded as legitimate in the matter under consideration. The measures shall be reviewed within a reasonable period of time, depending on the nature of the risk to life or health identified and the type of scientific information needed to clarify the scientific uncertainty and to conduct a more comprehensive risk assessment.'

If no scientific uncertainty persists, that is to say if risk assessment is conclusive, the conditions of Article 7 GFL are not met and the precautionary principle cannot be invoked.

4.6 Criteria

Criteria for authorisation differ between premarket approval schemes. See Diagram 4 for an illustration. The wording differs, but the three main criteria seem to be similar. The product must not be unsafe, must not be nutritionally disadvantageous and must not mislead the consumer. The three major criteria are all negative e.g. they are about what the food must *not* do. The text of the law seems to leave little leeway for balancing risks and benefits. Possible benefits are not mentioned as criteria, therefore it seems unlikely that benefits will be allowed to outweigh disadvantages.[72]

4.7 Case study novel foods

4.7.1 Introduction

To gain further insight into the effects of premarket approval schemes on food businesses' competitiveness, here below we will present a case study on novel foods. Among the different schemes, for several reasons we have chosen novel

---

[72] The coming Regulation on food additives (Article 6(2) Regulation 1333/2008) adds to the existing criterion of technological need, a requirement of advantages and benefits for consumers. It does not, however, open the option to balance risks and benefits.

Diagram 4. criteria for premarket approval.

	Additives (including sweeteners and colours)	Food supplements[1]	Novel foods	GM-food	Functional foods
Presence					
Technological need	X				
Scientific substantiation of claimed effect					X
Absence					
Safety hazard	X				
Danger for consumer			X		
Adverse effects on health		?		X	
Adverse effects on the environment				X	
Misleading consumer	X		X	X	X
Nutritional disadvantage			X	X	
Raise doubt about other foods					X
Encourage excessive consumption					X

[1] Directive 2002/46 only gives the positive list. It is silent on the procedure to update it and the criteria to be applied. The ECJ accepted this imperfection of the law in its ruling of 12 July 2005 (joined cases C-154/04 and C-155/04 consideration 85) deriving from the recitals that the only relevant criteria relate to the normal presence of the vitamin or mineral in the diet, its safety and its bioavailability.

foods. In the first food competitiveness study we collected data on novel foods and GMOs as representatives of the premarket approval schemes. Respondents indicated these as most restrictive for innovation. In many comments on EU premarket approval policy novel foods are addressed. They stand as *pars pro toto*. This is understandable as novel foods, along with genetically modified foods, is the most general category. Novelty is mainly related to the calendar date of market introduction not to function in (food) production or product. We have not chosen genetically modified foods because the discussion on this issue is obscured by strong emotions of political, ethical and other natures. They are also further away than novel foods from the dairy sector on which the second food competitiveness study focusses. The novel foods scheme has recently been evaluated and a proposal for improvements is on the table.

4.7.2 Background

On 14 February 1997, Regulation 258/97 concerning Novel Foods and Novel Food Ingredients, was published in the Official Journal. The regulation entered into force on 15 May 1997. Concerns about genetic modification were strong motivators to set up a system of premarket approval for innovative foods. Until 2004 this Regulation included genetically modified food. In 2004, genetically modified food became regulated by Regulation 1829/2003. This regulation set up an EU system to trace genetically modified organisms, introduce the labelling of genetically modified feed, reinforce the existing labelling rules for genetically modified food and establish an authorisation procedure for genetically modified organisms in food and feed and their deliberate release into the environment. Consequently, rules about genetically modified food were taken out of Regulation No 258/97; Articles 1(2)(a)(b), 3(3) and 9 were removed and Articles 3(2), 3(4), 8(1)(d) and 13 were modified. Currently Regulation 258/97 applies to novel foods that have not been subject to genetic modification.

4.7.3 Novelty

In Article 1, Regulation 258/97 defines its scope:

This Regulation shall apply to the placing on the market within the Community of foods and food ingredients which have not hitherto been used for human consumption to a significant degree within the Community and which fall under the following categories:
(c) foods and food ingredients with a new or intentionally modified primary molecular structure;
(d) foods and food ingredients consisting of or isolated from micro-organisms, fungi or algae;
(e) foods and food ingredients consisting of or isolated from plants and food ingredients isolated from animals, except for foods and food ingredients obtained by traditional propagating or breeding practices and having a history of safe food use;
(f) foods and food ingredients to which has been applied a production process not currently used, where that process gives rise to significant changes in the composition or structure of the foods or food ingredients which affect their nutritional value, metabolism or level of undesirable substances.

From our research as well as from the evaluation of the Novel Foods Regulation (hereafter: NFR) it follows that stakeholders find this scope difficult to deal with. Some aspects have become clear over time ('hitherto' means before 15 May 1997[73]), others remain obscure (in particular 'significant degree' and 'significant changes').

---

[73] According to an interviewee from DG Sanco, this date also applies to Member States who joined the EU later, bringing foods that established themselves on their market between 15 May 1997 and the date of accession within the ambit of the NFR. At least one example has been recorded.

Clarity on the scope of the NFR is important for at least three reasons. Food businesses need to know if the products they bring to the market need to be approved, enforcement officers need to be able to recognise unauthorised novel foods and administrative authorities need to know if they are competent to authorise a food presented as novel.

In itself, the use in legislation of open or otherwise vague concepts is not unusual, indeed it is unavoidable. Over time in their application to specific cases such norms usually acquire substance and a better delineated content. This raises the question why, after more than ten years of application, basic concepts in the NFR still remain unclarified. The answer to this riddle can be found, at least in part, in the way the NFR is being applied.

Decisions on applications for market authorisation for novel foods are normally[74] taken by the Commission. Even though the Commission's competence to approve depends on the applicability of the NFR and thus on the presence of a novel food, with only a few exceptions[75] in its decisions the Commission does not give an opinion let alone a reasoned opinion that the product at issue qualifies as a novel food. On this point the Commission is scarce in complying with the duty to give reasons as envisaged in Article 253 of the EC Treaty.[76] For this reason, the practice established by the European Commission has not made the contribution to clarifying the concept of novelty that could have been expected.

Decisions on novel foods applications fall short in stating reasons.

The first food competitiveness study relates the experience of interviewees that encountered closed doors when trying to discuss their products with the authorities

---

[74] The NFR applies a two stage procedure. The initial decision is taken at Member State level. If objections are raised – which is the case almost every time – competence shifts to the Commission.

[75] An exception is Commission Decision of 15 February 2002 authorising the placing on the market of coagulated potato proteins and hydrolysates thereof as novel food ingredients under Regulation 258/97 of the European Parliament and of the Council (2002/150/EC). Recital 1 of this decision reads: 'Whilst protein has been extracted from a number of plants to be used in foods, potato protein had not been on the market in the Community before Regulation 258/97 entered into force. Therefore, potato protein requires authorisation according to Article 1(2)(e) of the Regulation.' Even in this exception the most critical issue is not addressed: which are the considerations to regard an extraction from a traditional food as novel? Other exceptions are Decision 2001/17 refusing Nagai nuts and Decision 2001/721 on trehalose.

[76] Article 253 EC Treaty: Regulations, directives and *decisions* adopted jointly by the European Parliament and the Council, and such acts *adopted by* the Council or *the Commission*, shall *state the reasons* on which they are based and shall refer to any proposals or opinions which were required to be obtained pursuant to this Treaty. (Emphasis added). See also Article 41(2) 3rd indent of the Charter of Fundamental Rights of the European Union recognising the obligation of the administration to give reasons for its decisions.

in their Member State or in Brussels. Nevertheless a structure is in place to deal with such situations.

The formal procedure as envisaged in Article 1(3) NFR provides for the possibility to determine through comitology whether a type of food (or food ingredient) falls within the scope of Article 1(2) NFR, that is to say; is a novel food. In practice this procedure is not applied.[77]

As the initial stage of the novel foods procedure takes place at Member State level, it is also that level that is first confronted with the question regarding novelty. According to interviewees, the Member States have established a practice of reporting all potential novel foods that come to their attention through applications, or in a more informal way, to the Novel Foods Working Group CAFAB (Competent Authority Food Assessment Body). CAFAB falls under the Section General Food Law of the SCFCAH.[78]

CAFAB consists of national authorities. Sometimes, when technical questions are at issue, members bring support from national advisory bodies to the meetings. CAFAB members inform each other if a certain food was consumed in their country to a significant degree before 15 May 1997. Generally this is done by e-mail or during the meetings that take place three or four times a year. CAFAB tries to reach a common position on the basis of this information.

Every Member State seems to have, however, a different view on what 'used to a significant degree' means. Therefore in CAFAB the practice has developed that whenever a Member State representative declares that in her or his home country the product has been consumed at a significant degree before 15 May 1997, the other Member States will accept this verdict and the food is not considered novel.

Some examples of these views have been: 2 ha of agricultural produce of the product, general availability at supermarkets and in one case in 2001, the UK Food Standards Agency informed an importer that the plant Oca (*Oxalis tuberosa*) was not novel because in three books the use of Oca was mentioned in European gardens at the beginning of the nineteen century.[79] To interviewees, it is not clear how Member States check the presence of a product before 1997.

> Qualification of foods as novel is based on compromise rather than compliance with legislative criteria.

---

[77] See abbreviations.

[78] The Standing Committee on the Food Chain and Animal Health is subdivided in nine sections that cover particular aspects of food safety. To support these sections, Commission working groups like CAFAB have been formed to take care of the preparatory work.

[79] Hoogland, 2006.

Further, in general interviewees have the impression that CAFAB tends to give members and the chair their way. Thus, for example, if one member considers a food to be novel, it will be regarded as such. In practice, the verdict 'novel' seems to be inspired by safety concerns, not by any legal interpretation of the NFR. If any member has a safety concern, the conclusion tends to be that there has to be a risk assessment and that – therefore – the product is to be considered novel. Notwithstanding Article 9 GFL[80] no stakeholders[81] are consulted regarding the qualification of products as novel foods.

CAFAB keeps a list of all products it considers to be novel; the 'CAFAB list' or 'novel foods catalogue'. According to interviewees, this list is kept confidential. Initially this was confirmed in this research,[82] mid June 2008; however, DG Sanco changed its policy and published the Novel Foods Catalogue on the internet.[83]

Until this change of policy the CAFAB list could not – or could only to a very limited extent – contribute to developing the vague concepts of the NFR. By consequence, the onus to decide if a food is novel is on food business operators without much guidance to rely on. Interviewees believe that authorities readily and without forming a considered opinion of their own, regard as novel foods all foods that businesses present as such under the NFR.

The new policy could very well contribute to a more open debate with stakeholders on the concept of novelty in the NFR.

4.7.4 Dossier analysis

In the NFR case study three dossiers have been analysed. The analysis is included in annex 2 to this report. The three dossiers show strategic application by the businesses concerned of the novel foods scheme. In all three cases it would have been arguable that another regulatory frame (additives, processing aids) applies that takes precedence over the Novel Foods Regulation. In all three cases, the

---

[80] There shall be open and transparent public consultation, directly or through representative bodies, during the preparation, evaluation and revision of food law, except where the urgency of the matter does not allow it.

[81] The inclusion of a product in the list directly affects the business that brought the product to the attention of its national authorities, but also other businesses engaged in the same product.

[82] Given the experience from the interviews that urban legends go around regarding the content of food law, we decided to test this opinion from stakeholders. To this end on 5 May 2008 we submitted an application for the CAFAB list under Regulation 1049/2001. By letter of 6 June 2008 (SANCO/E4/AK/bs(2008) D/540255) the request was indeed refused by the Director-General Sanco. The letter explains that the CAFAB list is covered by one of the exceptions provided for by the policy relating to access to documents and that it cannot be made available. The exception is that the draft working document contains opinions for internal use which are part of deliberations and preliminary consultations. Its disclosure at this stage could seriously undermine the Commission decision-making process.

[83] Available at: http://ec.europa.eu/food/food/biotechnology/novelfood/index_en.htm. DG Sanco was so kind to inform the researchers of the new policy by e-mail of 19 June 2008.

competent authorities followed the qualification chosen by the applicant. It seems that the applicants value the protection a novel foods approval adds to patents. A novel foods approval grants a quasi-monopoly that is enforced under public law. Further the novel foods procedure does not involve the European Parliament, which may make the outcome of the procedure a little less unpredictable than an application for approval of an additive.

The dossiers also show pre-existing risk assessments either from a Member State or the Scientific Committee for Food predating the Novel Foods Regulation, or from outside the EU.

4.7.5 Enforcement

Interviewees believe that food safety inspectors are unable to recognise novel foods as such. Enforcement is limited to known novel foods. These are novel foods that have been approved (but are marketed by businesses that are not the authorisation holder), novel foods of which approval has been refused, novel foods that are in procedure and novel foods that have otherwise come to the attention of the authorities mainly because businesses have approached them to discuss their status (the CAFAB-list).

> Controls on non-approved novel foods seem limited to products on the CAFAB list.

Interviewed enforcement authorities have indicated they give limited priority to enforcement of the ban on unauthorised foods. By contrast to known pathogens the risk they pose is considered hypothetical. Interestingly the Dutch food safety authority (VWA) issued a risk assessment regarding unauthorised (GM) long-grain rice.[84] This seems to indicate that this authority is of the opinion that it is not entitled to enforce unless a risk to health has been established. Such an opinion would be in conformity with the general principle requiring scientific justification for limits placed on food businesses. It disregards, however, the shift in burden of proof based on the precautionary principle included in premarket approval requirements. Unapproved novel food is simply illegal and therefore can be made the subject of enforcement.

4.7.6 Business strategies

The available data show at least three, probably four, different business strategies with regard to the Novel Foods Regulation. Most businesses avoid all innovation

---

[84] RIVM, 2006.

that would bring them within the ambit of premarket approval requirements, some take on the required procedure, some do not avoid the innovation but try to stay out of the procedures and some just ignore the law.

To a certain extent these strategies can be explained by economic factors. Graham Brookes[85] analysed the economic impact of the novel foods approval procedures on the EU food sector. He found that it is fairly common for the costs associated with meeting regulatory requirements to be between € 0.3 million and € 4 million and that the considerable additional time taken to authorise novel foods in the EU adds an extra € 0.3 million to € 0.75 million per application. He found that the rate of return of the costs made on these investments would be 24%-25% if the procedure were to take 6 months. If delayed to 2.5 to 3 years then the rate decreases to 17%-18% and if it is extended to five years (60 months) it becomes negative as the rate is then 14.6% which is lower than 15%, the commonly used baseline for determining whether investments take place. The first food competitiveness study found an average of three years (between nine months and eight years[86]). Apparently at least some potential innovators fear not to get the desired return on investment and abstain from the innovation.

Avoidance
As has already been highlighted in Wijnands *et al.* (2007), one group of food businesses refrains from all forms of innovation that would bring them within the ambit of the NFR. In the NFR case study two interviewed businesses related that repeatedly (about twice a year) they pass by or abort projects that would risk bringing them within the ambit of premarket approval schemes.[87]

Monopolisation
There are, of course, businesses that do engage in premarket approval procedures. Beside a genuine desire to comply with legal requirements, different factors may contribute to explain this behaviour. In producer - customer relations an official proof of safety and/or marketability may be demanded.

Premarket approval schemes may provide competitive advantages. One of the larger businesses interviewed in the first food competitiveness study remarked that the legislation is throwing a quasi-monopoly into its lap. Requirements are

---

[85] Brookes, 2007.

[86] Wijnands *et al.*, 2007, p. 82.

[87] After this report was submitted to the European Commission, this finding was confirmed by further MSc-research at Wageningen University. Elisabeth Smith finalised a thesis on exotic novel foods. Among her interviewees was a consultant for a development NGO. This consultant related that she advises third world producers against attempts to exports fruits and vegetables to the EU market if they risk being considered novel foods and to focus on other markets instead such as the USA, Japan and Australia. Further, in general her research shows premarket approval requirements to constitute insurmountable barriers to producers in the third world wishing to enter the EU market.

perceived as so difficult to fulfil that most potential competitors do not even try. The findings from the three dossier analyses discussed in 4.7.4 and annex 2 show that some businesses actively choose the NFR-procedure even in cases where non-applicability of the novel foods procedure would have been arguable. The cases show businesses keen on investing in protection of inventions. The key players in all three cases acquired one or several patents relating to the food product concerned. Authorisation under the NFR – unlike for example additives – is exclusive. It protects the applicant with a monopoly in ways comparable to a patent. Unlike a patent, however, enforcement of this monopoly is not strictly a private law matter, but also a matter of general interest undertaken by public authorities on their own initiative. Premarket approval schemes shield the market in two ways. First, they limit competition in that only big or specialised companies can successfully follow the procedure. Second, in the case of exclusive authorisation the market is also reserved to the applicant after authorisation.

To put it bluntly, the premarket approval procedure can be used by businesses as an instrument to clear the market of competitors. It seems likely that SMEs will be among the first to be discouraged.

Premarket approval requirements can be used by businesses for monopolisation purposes.

A last reason why NFR approval is preferred by some businesses to additive approval, is the role of the European Parliament in approving additives. This makes the decision to such an extent political and subject to lobbying that the result is less predictable than under the NFR.

Circumvention
A third strategy does not avoid *innovation* but looks consciously or unconsciously[88] for ways to nevertheless remain out of the scope of approval schemes. In the NFR case study a food safety inspector remarked that many companies just place a product on the market and when confronted they act surprised.

Given the above observation that authorities do not recognise novel foods unless these foods have been brought to their attention, the decision and timing to contact authorities is considered crucial. Once a product (as such or resulting from a certain process) is considered novel long delays are suffered and products may need to be taken from the market. This is true for products that objectively

---

[88] Given the observation (in section 3.1) that knowledge of food law is limited in food businesses, it seems probable that some market parties simply are not aware that premarket approval requirements apply to them.

meet the definition of novel foods, but also for products for which authorities can be made to believe that they do. Interviewees believe that the latter is easily done. For this reason there is always the fear that some competitor may spoil the market by suggesting that a product (resulting from a certain process) may be novel. These businesses on the one hand build their case by collecting evidence and arguments in support of non-novelty and on the other hand try to convince competitors working along similar lines, not to contact the authorities.[89] This strategy encounters problems if no agreement is reached or maintained or if customers demand an official statement of non-novelty.

Infringement
Some situations are difficult to understand differently than as wilful infringements on the NFR. Probably the most notorious novel food for which market authorisation has explicitly been denied, is stevia.[90] Nevertheless this product is readily available on the EU market.[91]

4.7.7 Stacked risk assessments

The NFR case study shows repeated risk assessment. The dossier analysis provides examples of risk assessments conducted prior to the entry into force of the Novel Foods Regulation and in other parts of world (Australia, JECFA). In virtually all applications under the NFR a favourable risk assessment at Member State level is followed by a risk assessment at EU level. Finally, from the exclusive nature of authorisations under some of the premarket approval schemes, it follows that businesses that want to bring to the market a product already approved for another applicant also have to provide scientific proof of safety (or under the NFR of substantial equivalence).

If we accept the interpretation of the European Commission that premarket approval schemes are based on the precautionary principle, from Article 7 GFL and the case law on the precautionary principle it follows that they can only be applied in cases where scientific uncertainty exists. In cases where conclusive risk assessment is available, the shift in the burden of proof can no longer be based on the precautionary principle. In such situations compliance with Article 6 GFL and Article XX(b) GATT would require the burden of proof to return to the authorities opposing market entry of the food product concerned.

---

[89] This type of collusion vis-à-vis the authorities is not about market behaviour and therefore - most likely - outside the scope of Article 81 of the EC Treaty (ban on cartels).
[90] Decision 2000/196.
[91] See among others: http://www.steviamarkt.nl/wat_is_stevia.asp; http://www.steviamarkt.com/; http://www.uw-kapper.nl/overig/stevia/stevia.html.

> The precautionary principle cannot support a burden on businesses to prove the safety of a product when authorities already have a conclusive risk assessment at their disposal.

Member States' practice (in comitology) to require additional risk assessment at EU level in situations where conclusive risk assessment has been provided by the competent authority in another Member State goes against the principle of proportionality and against Article 6 GFL if it is not supported by science-based concerns about the validity of the risk assessment provided.

4.7.8 Consequences

From the above, a picture emerges of a regulatory framework that is being applied by the stakeholders concerned without much regard for the requirements of the law but rather with their own strategic aims in mind. In consequence the legal framework remains vague and elusive.

Authorities do not justify their exercise of powers derived from the NFR by substantiating that legal requirements are met, but compromise on NFR applicability if Member State representatives believe they see safety issues. Businesses avoid the area altogether, try to turn it to their advantage or play hide and seek. This is a picture where it seems likely that novel products can make it to the market without being recognised as such.

The total number of products and processes that, during the last ten years, have been assessed under the Novel Foods Regulation is very limited. Only a handful of products have been rejected.[92] These rejections show the effect of reversal of the burden of proof. The Commission does not base its rejection on positive findings of risk but only on the consideration that 'it has not been demonstrated that the product complies with the criteria laid down in Article 3(1) of the Regulation'. The rejected products, stevia and nangai nuts, are considered safe food in other parts of the world.[93]

> The available data do not show a contribution of the Novel Foods Regulation to food safety.

---

[92] Apart from stevia mentioned above, also nangai nuts (Decision 2001/17) and betaine (Decision 2005/580) have been rejected. In the initial risk assessment, stevia had received a negative opinion in Belgium, nangai nuts a positive opinion in France and betaine a positive opinion in Finland.
[93] For stevioside (the active substance in stevia) JECFA advises an ADI of 5 mg per day/kg bodyweight.

All this – the very limited number of rejected products, the reasons given for rejection, the fact that no safety concerns seem to be present outside the EU regarding the rejected products, the presence of rejected products on the EU market and the low priority given to enforcement – raises the question whether the NFR makes a serious contribution to food safety in the EU[94] and, even more, whether the contribution it makes justifies the burden placed on food businesses in the EU.

4.7.9 Legislative proposals

The Commission has introduced proposals with the Council and the European Parliament for a common procedure and for a new Novel Foods Regulation (hereafter: NNFR). The question arises to what extent these would lessen the problems indicated above and to what extent additional remedies are required.

Clarity of concept
While the concepts used in the NFR have been somewhat simplified in the proposed new Novel Food Regulation, the core of vagueness – the notion of 'significant degree' – remains in place.

Clarity of access
A critical issue identified in the first food competitiveness study is a lack of clarity concerning the applicable procedure. The recommendation has been made to introduce a fast track procedure to identify the appropriate procedure or to come to a negative clearance in the form of a decision that no premarket approval is needed. The coming common procedure (Regulation 1331/2008) would have been an ideal place for such a procedure. Just like the current Novel Foods Regulation (Article 1(3)) the NNFR includes a provision (Article 2(3)), that provides the possibility to answer by comitology the question as to whether a type of food falls within its scope. Similar provisions were initially proposed for the coming regulations on enzymes,[95] on flavourings[96] and on additives.[97] The proposed provisions were not ideal in that they addressed the question regarding the applicable framework not in general but regulation by regulation e.g. for each sectoral food law separately. To a certain extent the Commission and the SCFCAH would have been able to make up for this deficit if they would have been willing to address open questions on applicability under these provisions simultaneously. Therefore the proposed provisions would have been a step in the right direction. For reasons that are not

---

[94] Two arguments are conceivable to justify the system. One is the expectation that it will make a serious contribution in the future. The other is that the number of rejections is so low because businesses will not try to gain approval for unsafe products that otherwise they would have brought to the market. These arguments are not, however, substantiated by empirical data.
[95] COM(2007) 670, Article 2(5), now Regulation 1332/2008
[96] COM(2007) 671, Article 2(3), now Regulation 1334/2008.
[97] COM(2007) 673, Article 2(5), now Regulation 1333/2008.

apparent from the published documents,[98] the proposed provisions have been deleted from the final texts.

The opportunity to create a simple and fast procedure to identify the applicable procedure, therefore has been bypassed. Not even the initially envisaged improvement on the current situation has been achieved.

Reversal of the burden of proof
As discussed above, premarket approval schemes reverse the burden of proof with regard to the safety of food products. It is not self-evident that such reversal is compatible with the WTO SPS Agreement. Article 5 SPS Agreement requires sanitary measures to be based on risk assessment. The NNFR, just as the NFR, regards novel foods as *a priori* unsafe until proven safe. No risk assessment underlies this presumption of lack of safety.[99]

Members of the WTO are beholden to notify their sanitary measures and to justify them when challenged by other members. The European Commission is said by stakeholders to have circumvented this obligation by notifying the NNFR under the TBT Agreement (technical barriers to trade) and not under the SPS Agreement.[100] This is a disturbing situation. The position that premarket approval schemes demanding proof of food safety are not sanitary (but technical) measures does not seem to be tenable. In this way the EU calls down on itself the suspicion that it is unwilling and maybe even unable to justify its measures.

The EU should notify the NNFR under the SPS Agreement.

De-politicisation
The first food competitiveness study and also the current study identified the political character of the approval procedure to cause problems. The first food

---

[98] The European Parliament proposed amendments to the effect that the regulatory procedure with scrutiny would be applicable, not to delete the entire provisions.

[99] Indeed Article 6 GFL notwithstanding (this article requires food law to be based on risk analysis and to take into account in particular the opinions of EFSA), no indication is available that EFSA's opinion has been asked prior to submitting the proposal or that it is based in some other way on risk analysis.

[100] Indeed on 14 March 2008 a notification has been filed with reference number; G/TBT/N/ EEC/188. See: http://useu.usmission.gov/agri/WTOnotif.html. In its summary report of the meeting of 8-9 October 2008 (G/SPS/R/53 of 22 December 2008, p. 6) the WTO Committee on Sanitary and Phytosanitary Measures, remarks in this respect: 'The representative of the European Communities noted that in this specific case, the legal advice had been to only notify the proposed revision to the TBT Committee since it covered approval procedures for Novel Foods in general. But, this did not preclude that the issue could be discussed at the SPS Committee. In response to his request, the Secretariat clarified that it generally recommended that any draft regulations with any SPS content should also be notified to the SPS Committee, even if these regulations were also notified to the TBT Committee.'

competitiveness study brought to light that stakeholders experienced political decision-making as being unpredictable in outcome. The current study goes further to explaining this situation by showing that this decision-making does not always remain within the limits of the law. This is understandable as the natural function of politics is to set rather than to apply the law.

The proposals replace the legislative procedure for updating the additives directives by the comitology procedure with scrutiny.

This contributes a little to replacing political decision-making by administrative decision-making. However as we have seen above, comitology has become very political as well. Also transparency is a sensitive issue in comitology. Insofar as it is not considered possible to set up an entirely administrative procedure without comitology, the Member States' representatives can contribute to improvement of the procedure by complying with the legal framework.

Deadlines
The long duration of approval procedures is burdensome on businesses. The proposed common procedure sets some deadlines, but leaves much of the timing open. For an overview see Diagram 5. The proposed regulation does not make it possible to estimate the time involved in the procedure. It can be over two years without any extension or surpassing of deadlines. As a consequence, the NNFR holds no promise for a reduction of the average time involved in the procedure.

Diagram 5. Timeframe in NNFR.

Applications must me addressed to the Commission	
Acknowledgement receipt (Art. 4)	14 working days (three weeks)
Opinion EFSA (Art. 5)	9 months
The nature of the matter so justifies (Art. 10)	extension PM
EFSA requires additional information (Art. 6)	extension PM
Commission submits draft regulation (Art. 7)	9 months
The nature of the matter so justifies (Art. 10)	extension PM
Commission requires additional information (Art. 8)	extension PM
Opinion SCFCAH (Art. 5a(2) Dec. 1999/468)	to be decided by its chair
A Commission concurs with SCFCAH (Art. 5a(3) Dec. 1999/468)	
Commission proposal to EP and Council	without delay
Opposition EP or Council	2 months
B Commission does not concur with SCFCAH (Art. 5a(4) Dec. 1999/468)	
Commission proposal to EP and Council	without delay
Councils acts	2 months
Decision by EP	4 months

Stakeholders fear that authorities consider surpassing deadlines an option. For this reason the first food competitiveness study recommended the introduction of fatal deadlines. The only deadlines in the proposed procedure that have a fatal character are the ones applying to the Council and European Parliament. For the Commission, EFSA and the SCFCAH, surpassing deadlines is without sanctions. This means that businesses depend on their goodwill to act exclusively on the basis of the law, respect the conditions of the law and refrain from going beyond the limits of the law.

It must be concluded that the proposal does not provide businesses with legal certainty regarding the time involved in the procedure.

Mutual recognition
In the first food competitiveness study investment was recommended in (mutual) recognition of risk assessment executed by competent risk assessment bodies such as FDA and JECFA.[101] The current study identified that as premarket approval is considered by the Commission to be based on the precautionary principle, the shift of the burden of scientific proof of safety cannot be justified in situations where conclusive risk assessment is available. Also for this reason it is necessary that a position be taken regarding pre-existing risk assessment. No provision to this effect is included in the proposals.

4.7.10 Concluding remarks

The legitimate aim of premarket approval schemes is to protect food safety in the EU. Based on the precautionary principle such schemes shift the burden of proof from authorities to businesses.

The schemes need clarification. The Commission has taken an important step in this direction by publishing the novel foods catalogue. Further steps are needed.

Proposed new legislation provides some improvements as well but does not fully comply with the precautionary principle.

---

[101] A first step in this direction has been taken in Regulation 470/2009. This regulation on residue limits for veterinary drugs opens the possibility to base EU limits on decisions of the Codex Alimentarius, which in turn are based on risk assessments by JECFA, JMPR and JEMRA. The regulation states in Article 14 (3)(b): 'A maximum residue limit shall be laid down where it appears necessary for the protection of human health: pursuant to a decision of the Codex Alimentarius Commission, without objection from the Community Delegation, in favour of a maximum residue limit for a pharmacologically active substance intended for use in a veterinary medicinal product, provided that the scientific data taken into consideration have been made available to the Community Delegation prior to the decision of the Codex Alimentarius Commission. In this case, an additional assessment by the Agency shall not be required.'

5. Food safety targets

In EU food law a fundamental norm applies that food may not be brought to the market if it is unsafe.[102] In general it is left to the responsibility of food businesses to determine what, from a scientific point of view, this means in practice.

For an increasing number of substances and microorganisms, legislation sets specific limits distinguishing what is considered safe from what is considered unsafe. In principle such legislation can be seen as helpful to businesses as it increases legal certainty. The onus to distinguish safe from unsafe is no longer on them with regard to the substances concerned.[103] A downside to such legislation is that it fixes in law, scientific understanding as it stands at a certain moment in time and thus may easily become outdated in the light of scientific progress.

If targets are being set at a very strict level that is difficult to achieve under normal conditions of production, they may constitute barriers to market entry. Such barriers are justified if they are necessary to protect public health.

In setting the target, toxicological and other scientific tests play an important role. Data from toxicological tests on the contaminant (like residues of a pesticide or veterinary drugs) allow for the fixing of an 'acceptable daily intake' (ADI). Usually this involves finding the highest dose that would produce no adverse effects over a lifetime (chronic) exposure period and then applying appropriate safety factors. From animal testing the NOAEL is derived; the 'No observed adverse effects level'. The quantity of contaminant represented by the NOAEL is twice divided by 10 to specify the ADI; once to account for the difference in sensitivity between species and once to account for the difference in sensitivity between individuals. As a consequence the ADI is generally 1/100 of the NOAEL. In the case of products that need premarket approval like pesticides and veterinary drugs, the ADI is compared with the residue level resulting if the product is used in a proper way, that is to say applying recognised good practices (good agricultural practices: GAP). Where this level is higher than the ADI, the product is not approved.

The most strict level conceivable is zero. Several substances and organisms may not be present in food at all ('zero-tolerance'). In practice 'zero' is the detection limit. With progress of technology ever smaller contaminations can be detected. Such developments risk taking the zero-norm *ad absurdum*.

---

[102] Article 14 regulation 178/2002.
[103] Dutch case law provides an example (Rechtbank - district court - Leeuwarden 28 May 2004, Gerechtshof - court of appeal - Leeuwarden 30 November 2006) where the presence in a food of a level of contamination not conforming to the most scientific opinion (but to older ones) was considered a criminal offence.

In the journal 'Trends in Food Science & Technology',[104] Huub Lelieveld and Larry Keener on behalf of GHI[105] call on scientists to become involved in discussions on harmonisation of food regulations and legislation. To convince their colleagues of the urgency of the engagement of science in food law, they relate the following:

> *'In 2001, the EU decided to destroy a large amount of fish containing minute amounts of chloramphenicol. Chloramphenicol is an antibiotic produced by Streptomyces venezuelae that is frequently prescribed for humans and other mammals. On 28 September 2006, the European Court of Justice, considering that zero-tolerance applies to furazolidone and chloramphenicol, ruled that EU countries must seize and destroy meat containing such substances, even if containing just ppbs.'*

Apparently, from the perspective of these authors, the application of zero-tolerance in a situation as described is absurd from a scientific point of view. At levels as at issue here, proven presence of a substance can no longer be seen as proof of a food safety risk (or even of use of illegal products). When due to technological developments zero-tolerance loses meaning, the legislator should take the responsibility on the basis of sound risk assessment to redefine 'zero' by setting a specific – presumably very small – detectable limit.

Zero tolerance applies for example to the veterinary drugs included in Annex IV of Regulation 2377/90.[106] In situations where it appears from risk assessment that a maximum residue limit cannot be established in respect of a pharmacologically active substance used in veterinary medicinal products because residues of the substances concerned, at whatever limit, in foodstuffs of animal origin constitute a hazard to the health of the consumer, that substance shall be included in a list in Annex IV. The administration of substances listed in Annex IV to food-producing animals is prohibited throughout the Community.[107] To substances not listed in Annex I, II or III, zero tolerance applies.

Annex IV holds substances like: *Aristolochia* spp. and preparations thereof, chloramphenicol, chloroform, chlorpromazine, colchicine, dapsone, dimetridazole, metronidazole, nitrofurans (including furazolidone) and ronidazole. The European Court of Justice has ruled in the case referred to above that if in controls an Annex-IV-substance is found (in that case 49 ppb and 2.7 ppb furazolidone and 1.4 ppb chloramphenicol), seizure and destruction of the food in question is

---

[104] Lelieveld and Keener, 2007.

[105] The Global Harmonization Initiative.

[106] Regulation 2377/90 of 26 June 1990 laying down a Community procedure for the establishment of maximum residue limits of veterinary medicinal products in foodstuffs of animal origin (OJ L 224 18.8.1990). This regulation is being replaced by Regulation 470/2009. The Annex-IV-approach is continued under the new legislation.

[107] Article 5 Regulation 2377/90.

imperative – no matter how minute the trace.[108] It is not allowed to return the product to its sender.

Zero-tolerance leads to the destruction of food if ppbs are detected.

On this last point an improvement has been made in Regulation 882/2004 on official controls.[109]

We interviewed the first author of the article in Trends in Food Science & Technology on this issue.[110] He pointed out that half of the drugs listed in Annex IV are antibiotics produced by micro-organisms (and thus) occurring naturally. Over the last twenty years precision of measurement techniques have improved by a factor of 1,000,000 This means that the norm 'zero' is now a millionth of what it was twenty years ago. It is inconceivable that the risk judgement that a maximum residue limit cannot be established is based on an impossibility of establishing a NOAEL. In other words that adverse effects have been observed of a dose in parts per billion (one single molecule among 1,000,000,000 others; one cube of sugar in Lake Maggiore). If it does not mean that such observations have been made, what does it mean? In some cases it may mean that there was not sufficient commercial interest for the producer of the drug or pesticide to support the product with tests showing a NOAEL.

The interviewee pointed out that the antibiotic at issue in the Court ruling (chloramphenicol) is administered for medicinal purposes to humans including babies at a dose of 25 mg per kg body weight. To match this therapeutic dose – apparently considered sufficiently safe – a baby would have to eat 500 kg per day of the 'contaminated' food at issue, an adult 35 tonnes. The interviewee considers the destruction of food posing no health risk that any scientist would endorse as highly unethical.

A position taken by the EC in the Codex Committee on Residues of Veterinary Drugs in Foods, explains how the 'zero' level is connected to a prohibition to administer the product. '*In the European Community substances intended for use*

---

[108] ECJ 28 September 2006 cases C-129/05 and C-130/05, NV Raverco (C-129/05), Coxon & Chatterton Ltd (C-130/05) v. Minister van Landbouw, Natuur en Voedselkwaliteit (NL).
[109] Article 21 of this regulation allows for re-dispatch of consignments not complying with food law if certain conditions are met. Article 12 GFL goes in the same direction. The food may not be injurious to health. It remains to be seen how this concept will be applied in relation to zero-tolerance norms.
[110] The interviewee shared the draft report of the interview with other scientists interested in GHI. Their reactions show support. This indicates that the text presented in this section is more than a single man's opinion. As the reaction comes from people who have not been selected randomly, it does not indicate, however, that the opinion is statistically relevant.

in food producing animals have to be evaluated with regard to the safety of residues. If the assessment indicates that residues of the substance concerned constitute a hazard to the health of the consumer, that substance will be prohibited or its use restricted in food producing animals. Substances prohibited because available data suggest that their use in food producing animals is generally unsafe.[111]

Zero-tolerance may result from other sources than scientifically-established safety levels.

Apparently the zero-level in cases such as this does not represent a risk assessment positively finding *risks*[112] at the most minute of levels, but the unapproved *status* of the product. In such a situation, however, applying zero-tolerance is no longer a risk management instrument but a sanctioning mechanism to deal with the use of unapproved substances. Such practices near abuse of risk management for other objectives. If ever such practice is acceptable, it certainly is not with regard to contaminations occurring in nature.[113]

The European Commission is oversimplifying the matter, when it takes the following position:

> *'Better testing techniques are not a problem per se – indeed they are welcome, but they can cause practical difficulties for some third countries, and in particular some developing countries. For example we can now detect routinely chloramphenicol at levels as low as 3 parts per 10 billion – some 3-15 times more sensitive than even one year ago. And sensitive tests are now available for metabolites of nitrofurans – a recent development. Developing countries often lack the equipment and the trained staff to enable testing at such levels. The way to address this issue is to encourage better control measures and improved infrastructure to bring developing countries up to EU standards to the benefit of all. The European Commission is assisting such efforts through trade related technical assistance.'*[114]

---

[111] European Community 2005.

[112] It should be noted that in the case of antibiotics two different types of risks are relevant: direct health risks to the consumer and the risk that pathogens develop resistance against the antibiotic.

[113] Another scientist pointed out by e-mail that medicines used in human medication have been proven through sampling to end up in the aquatic environment by way of the sewage system. 'Although a direct correlation with occurrence of residues through human therapeutic use and presence of these antibiotic residues in aquatic animals has never been proven, the essence is that it also cannot be ruled out. The direct assumption, that presence of these residues constitute evidence of fraudulent production of the (sea)food, has therefore become irrational at ppb level.'

[114] European Commission, 2003.

Preferable – indeed necessary – is the approach chosen in the new legislation on pesticides. Regulation 396/2005[115] sets a general maximum pesticide residue level (MRL) in foodstuffs of 0.01 mg/kg.[116] This general level is applicable 'by default', i.e. in all cases where an MRL has not been specifically set for a product - food combination. Such a default level should be set for *all* contaminants. Lower levels should be set only on the basis of scientific risk assessment and never for reasons like lack of producer support or coercion of producers in – as well as outside – the EU into refraining from the use of products not conforming to EU standards.

Default levels are preferable to zero-tolerance.

In cases where Article 5 of Regulation 2377/90[117] applies to the letter, in the sense that risk assessment does not yield a safety level, apparently we are dealing with situations covered by Article 7 GFL. Scientific uncertainty persists. The setting of a zero level in the extreme meaning of the word is, therefore, a precautionary measure. Article 7 GFL allows for provisional measures only. The authority taking such measures is beholden to ensure that additional risk assessment is undertaken.

Zero limits as precautionary measures should always be followed by further risk assessment.

The interviewee addressed another subject regarding safety assessment and the setting of food safety limits. Currently animal testing is the regulatory standard for safety assessment. Scientists however agree that this type of testing holds limited relevance for judging the safety of a product for humans. Products safe for humans may be considered hazardous while hazardous products can be overlooked if they are safe for the species of animal used. For the last ten years, thanks to research funded by the EU,[118] new technologies have become available that yield results with considerable higher human relevance. HEPADNA is based on human HEP G2 cell lines (from the liver) that have retained most of their functionality and for this reason represent better human models than animals do.[119]

---

[115] Regulation 396/2005 of the European Parliament and of the Council of 23 February 2005 on maximum residue levels of pesticides in products of plant and animal origin, OJ L 70/1 16.3.2005.

[116] A comparable approach is found in regulation 1829/2003. Article 47 states that - if certain conditions are met - the presence of an unapproved GMO in a food shall not be considered to be in breach of the regulation, if the proportion is not higher than 0.5%.

[117] Where it appears that a maximum residue limit cannot be established in respect of a pharmacologically active substance used in veterinary medicinal products because residues of the substances concerned, at whatever limit, in foodstuffs of animal origin constitute a hazard to the health of the consumer (...).

[118] Available at: http://hepadna.com/.

[119] There would also be animal welfare benefits.

The best available human models should be recognised for risk assessment.

The interviewee advises:
- introduction of a default level for all forms of contaminants;
- setting stricter levels than the default level exclusively on the basis of scientific risk assessment;
- replacing animal testing by HEPADNA testing as the preferred mode of risk assessment in the regulatory framework on food safety.

6. Food handling and controls

6.1 General

Process requirements on traceability and hygiene are said to push SMEs out of business. What is the situation?

The French cheese case study and the Alpeggio case study show that virtually all interviewees consider food hygiene to be very important. An interesting observation from the French cheese case study is that good hygiene practices result in a premium. Milk low in microbiological contamination is of high value for the dairy industry, in particular for producers of raw-milk-based cheese. The price paid by processors varies with the microbiological cell count. In this sense EU hygiene legislation is believed to increase rather than threaten competitiveness of the EU dairy sector. The interviewed processors fear that when the quota system ends, dairy farmers will focus on quantity rather than quality resulting in reduced availability of the high quality inputs they need. Already larger processors are insisting on pasteurisation rather than quality control.[120]

6.2 Flexibility

The hygiene package aims to be flexible in dealing with traditional and disadvantaged production. Is this flexibility effective?

The concept of flexibility has its current legal base in Article 13(2), (3) and (4) of Regulation 852/2004 as elaborated and expanded in Commission Regulation 2074/2005. Member States may take measures enabling the continued use of traditional methods, at any of the stages of production, processing or distribution of food or accommodating the needs of food businesses situated in regions that are subject to special geographical constraints. Further, in general they are entitled to adapt the EU requirements regarding the construction, layout and equipment of establishments.

The Alpeggio case study shows that the derogations needed in practice to sustain traditional ways of cheese production, do not go beyond the options available in EU Hygiene legislation. Whether the derogations are granted, however, depends entirely on the Member States concerned. In Italy this competence is decentralised.

---

[120] This finding from the French cheese case study is confirmed in the press. See for examples: http://www.independent.co.uk/news/world/europe/cheeseknives-at-dawn-in-camembert-457168.html; http://www.guardian.co.uk/world/2008/apr/18/france; http://www.timesonline.co.uk/tol/news/world/europe/article1459347.ece; http://blogs.telegraph.co.uk/jon__doust/blog/2008/04/10/small_war_in_camembert_country; http://www.news24.com/News24/World/News/0,,2-10-1462_2125070,00.html; http://www.expatica.com/fr/articles/news/_Real-Camembert_-wins-war-against-supermarket-variety.html.

Each region has a different orientation. Thus Lombardia and Valle d'Aosta on the basis of the hygiene package enacted provisions to guarantee a certain flexibility to micro establishments whereas Veneto provided a sub-authorisation for those alpeggi that cannot meet the criteria of Regulation 852/2004, rather than applying flexibility as given in Article 13(2) of this regulation. Piemonte did not take any position despite the significant number of micro-dairies in its territory. The biggest discrepancy was observed during the interviews of a producer in the Province of Sondrio, Lombardia, allowed to use tents as provisional premises, and a producer in the Province of Vercelli, Piemonte, who was instructed to plaster the dry-stone walls of the ripening room.

Not only does the application of EU law appear to depend on the Region's orientation, interviewees also observed variability of discretion between officials of different local authorities in the same Region and even within the same local authority. The impression resulting from this case study is that the flexibility needed by micro and traditional enterprises is actually present in the EU hygiene package but that businesses and Member States are not sufficiently aware of this. For this reason the burden perceived from food hygiene legislation is higher than is justified by the content of the law.

6.3 Record keeping

Interviewees in the Alpeggio case study and in the French cheese case study consider record keeping an embarrassment. They call it boring and a loss of valuable time. No doubt these interviewees will welcome the proposal to exempt them from these requirements in the context of Regulation 852/2004 on HACCP.[121] However, UEAPME, the organisation representing small and medium-sized enterprises in Brussels, has a different opinion. In a letter to the European Commission it opposed this proposal stating that small businesses are fully capable of responsible production.

6.4 Inspection costs

Stakeholders have brought another issue to our attention. Official controls are organised on a risk basis.[122] In some Member States inspectors consider traditional methods – the use of raw milk in particular – to be high risk. For this reason traditional producers face a high frequency in controls. Member States may –

---

[121] COM(2007) 23, priority area food safety; p. 21. See also: COM(2007) 90 final. Holding the provision (Article 2): In Article 5(3) of Regulation 852/2004, the following sentence is added: 'Without prejudice to the other requirements of this Regulation, paragraph 1 shall not apply to businesses which are micro-enterprises within the meaning of Commission Recommendation 2003/361/EC of 6 May 2003 and the activities of which consist predominantly in the direct sale of food to the final consumer.'
[122] Article 3 Regulation 882/2004.

sometimes must – collect fees to cover the costs of controls.[123] In this way costs of controls on traditional methods are being raised to such a level in some Member States that smaller businesses have had to abandon such methods.

6.5 Concluding remarks

Food businesses perceive threats from food law to traditional production methods. Confronting the legislation with the practical needs of traditional producers, shows that all the needed exceptions can be provided by the Member States. The possibilities are not used however to the extent possible. On top of this official controls are organised and charged in ways that put small traditional producers at a disadvantage.

---

[123] Regulation 882/2004; Chapter VI Financing of Official Controls, Articles 26-29.

7. Labelling

7.1 Current situation

Experience (in terms of attendance) from meetings of food law associations in Germany, the Netherlands and at EU level, with courses and workshops shows that labelling is considered the single most important legal issue for food businesses. More than is the case with other parts of food legislation, businesses seem to be aware of the national legislation on food labelling (based on the General labelling directive 2000/13/EC).

The first food competitiveness study indicates that labelling is in particular a critical issue for SMEs. Also with regard to labelling, businesses appreciate the level playing field the EU (food) law has created within the EU (now) 27.

The biggest problem is the costs in particular related to small scale production and seasonal or other variations in the composition of products. Labelling requirements sit better with standardised industrial production than with artisanal production. Unlike premarket approval, labelling is an area whose burdens cannot be avoided by staying away from it. Labelling requirements apply to all food businesses. This may explain the surprising finding from the survey in the second food competitiveness study that labelling is ranked highest (but not very high: N 24, mean 4.21 on a 1 low to 7 high scale) as an area of food law restricting innovation.

Labelling is a prime target for inspection and therefore an area where relatively often adjustments have to be made to indulge the inspector. A perceived problem in labelling legislation is that 'there always is something else'. This means that if a business has a label complying with the general requirements, in its perception there may still pop up some specific requirement in some exotic piece of legislation.

7.2 Proposal for a new regulation

DG Sanco has undertaken a project called 'Labelling: competitiveness, consumer information and better regulation for the EU'.[124] This project resulted on 30 January 2008 in a 'Proposal for a Regulation of the European Parliament and of the Council on the provision of food information to consumers'.[125] Does the proposed new regulation sufficiently address businesses' worries?

---

[124] In February 2006 DG Sanco published a consultative document in preparation of new food labelling legislation under the title 'Labelling: competitiveness, consumer information and better regulation for the EU' (DG Sanco 2006).
[125] COM(2008) 40 final.

The proposal has been addressed in numerous meetings of experts and business representatives. For the current study opinions have been gathered at a workshop organised by the European Food Law Association (EFLA) in Brussels on 11 February 2008, the CIES[126] food safety conference on 14 and 15 February 2008 in Amsterdam, a meeting of the UEAPME[127] working group on foodstuffs on 28 February 2008 in Brussels, the Lebensmittelrechtstage, Wiessbaden Germany 6 and 7 March 2008, a workshop on the new labelling proposal organised by the Dutch Food Law Association (NVLR[128]) on 12 March 2008, the MLK[129] '"Gesunde" Lebensmittel-Kennzeichnung. Was ist das und gibt es sie überhaupt?' Münster Germany 31 March 2008 and at the seminar Regulating Food Safety and Environmental Protection: legal challenges, at the University of Copenhagen 19 and 20 May 2008. Even taking account of the fact that experts meeting at fora as the above are more inclined to voice criticism than approval, the overall reception of the proposal was highly critical.[130]

Stakeholders criticise the new labelling proposal as being insufficiently balanced.

The new proposal is presented as a fast track project within better regulation and is meant to support business competitiveness. Nevertheless, the substantive changes seem limited compared to the major effort that is being placed upon businesses to readjust to the legislation. The project holds four key issues: mandatory front-of-pack nutrition labelling, mandatory font size for all mandatory information, additional legislative powers for the Commission and voluntary national schemes. According to experts, for the two latter no convincing need has been shown, while for the two former just two articles would be sufficient to achieve the desired result.

Some elements – in particular 3 mm minimum font size – are perceived as unnecessarily burdensome by food businesses. Opinions about nutrition labelling and its usefulness are mixed. The general feeling in the food sector seems to be that the results from the proposed new regulation come at too high a price.

---

[126] Comité International d'Entreprises à Succursales (CIES – International Committee of Food Retail Chains) http://www.ciesnet.com/.
[127] UEAPME: union européenne de l'artisanat et des petites et moyennes entreprises, www.ueapme. com.
[128] NVLR: Nederlandse Vereniging voor Levensmiddelenrecht, www.nvlr.nl.
[129] MLK: Münster´sches Lebensmittelrechts-Kolloquium.
[130] First comments in literature are critical as well. See for example: Grit and Tiesinga, 2008; Hagenmeyer, 2008; Kozlovski and Van der Kroon, 2008; Mettke, 2008.

7.3 Stability

Among the main findings in the first food competitiveness study is businesses' worry about a lack in stability in EU food law. Businesses want to see changes in legislation limited to situations where such changes are necessary and bring substantial improvements.[131]

The proposal brings substantial changes in the form of the law. Harmonised national legislation is replaced by a regulation at EU level. This is a commercial blessing for those offering food law education to the market (including the authors of this report), but a burden on food businesses. The explanatory memorandum takes the opposite view stating that a Regulation provides a consistent approach for industry to follow and reduces the administrative burden as they need not familiarise themselves with the individual regulations in the Member States. This is true only for businesses accessing new markets in other Member States. It is not true for businesses continuing their activities on a mainly national market as is the case with most SMEs. The preceding statement in the explanatory memorandum is a bold one: 'A Directive would have led to an inconsistent approach in the Community leading to uncertainty for both consumers and the industry.' Stakeholders fully agree that inconsistency and uncertainty should be avoided at all costs. This is precisely the reason why they deplore the call the proposal makes upon the Member States to develop national schemes (chapter VII of the proposed regulation).

Stakeholders oppose the provisions on national schemes.

According to stakeholders, it overestimates national inspectors' ability to differentiate between binding law and non-binding national schemes that should not constitute a barrier to international trade. In practice the non-binding but very much cherished national schemes on food labelling based on composition (like the German and Austrian Lebensmittelbuch) constitute high barriers to intra-Community trade. Everybody knows that their application to imported products is inconsistent with EU law. Interviewees say that inspectors openly acknowledge that the business will win the case (several years later) if they take it to court, but that compliance is the only viable option for market entry in the short term. On the basis of such experience interviewees fiercely oppose the invitation to create more national schemes. The proposal is expected to bring a decrease instead of an increase of harmonisation of food labelling law in the EU.

---

[131] Wijnands *et al.*, 2007, p. 67.

7.4 Consolidation

The proposal subscribes to the need to codify EU food labelling legislation. To this effect the explanatory memorandum states:

> 'The proposal modernises, simplifies and clarifies the current food labelling scene. In particular:
>
> Recasting of the different horizontal labelling provisions. The merging of those texts (directives) into a single piece of legislation (regulation) will maximise synergies and increase the clarity and consistency of Community rules. This is a powerful simplification method that should provide economic operators and enforcement authorities with a clearer and more streamlined regulatory framework.'

Indeed opinions of experts and business stakeholders concur that merging all food labelling provisions into one single text would be a great improvement. Despite the high expectations that the quote above might raise, the proposal by no means approaches a full codification. The project set out on the wrong foot. Already in the consultative document,[132] the Commission presented full codification as unattainable:

> '18. What is the most appropriate legislative instrument *to implement these laws more homogenously in the European market (Member States have regularly spoken in favour of a regulation instead of a directive) and* how should the labelling provisions be brought together? *It is absolutely true that labelling or labelling-related provisions are included in many pieces of legislation, but this is the consequence of the widely used rule of* Lex generalis *and* Lex specialis. *Common labelling requirements applicable to all foodstuffs are laid down in horizontal legislation (Directive 2000/13/EC and related texts), whilst specific provisions, because of specific needs to informing consumers, are included in vertical legislation, as a result of specific composition or quality standards to which they are closely linked. The same structure is used in Member States national legislation as well as in international standards of Codex alimentarius.*
>
> *19. It would be very difficult to compile all specific information requirements, applicable from fish to chocolate for example, in the same legislative package. Furthermore, such a work could result in new inconsistencies and would neither be easily usable by operators nor manageable by Public Authorities. It would perhaps be more feasible and hopeful to* recast all horizontal provisions in a single proposal. *Such an approach should also seek to present, simplify and clarify the provisions currently spread across these texts, which could be brought together in an annex.'[133]*

The inconsistencies referred to in the quote above will not go away if vertical food labelling requirements are being kept out of the codification (as are the very important horizontal nutrition and health claims Regulation; the regulations on

---

[132] DG Sanco 2006.
[133] Emphasis in original.

protected designations, GMOs, novel foods, etc.). They will be left to the food businesses to resolve.

The very least that should be done is add an exhaustive annex listing *all* other food labelling provisions. This may have been what was meant by the last word in the quote above. None of the 13 annexes to the proposal provide this information however. In particular article 18 of the proposal in conjunction with Article 2(2) (1) requiring the legal name of the product to appear on the label would be well served with an annex listing all Community provisions prescribing such names. The authors of this report would like to add to the stakeholders' opinions that it would be befitting of the Community's commitment to the development of international standards[134] to explicitly include the names coined in Codex Alimentarius standards as legal names (but not prescribed names) in EU food labelling law.

Stakeholders support full codification of EU food labelling law.

7.5 Further proliferation

A further threat the proposal poses to a simple, coherent and accessible structure is the abundance of delegations to the Commission of powers to create additional provisions complicating the structure.

Twenty-three times the proposal grants the power to the Commission to amend it, each time repeating the line: those measures designed to amend non-essential elements of this Regulation by supplementing it shall be adopted, in accordance with the regulatory procedure with scrutiny referred to in Article 49(...).

7.6 Stakeholders' views in short

Interviewees consider various aspects of the proposal to be positive. In particular the step that is being taken towards codification of food labelling law and the clarification of some concepts like the origin of products and the values to be applied in nutrition labelling. Not positive in their view are the mandatory minimum font size of 3mm for all mandatory information on the label. According to interviewees this is not practicable.[135] The invitation for differing national schemes and the powers for the Commission to create hordes of amendments are not appreciated. The fact that the Commission shied away from full codification is not seen as very

---

[134] Articles 5(3) and 13 GFL.

[135] The exception for labels smaller than 10 cm^2 is relevant only for packages the size of a matchbox. One interviewee demonstrated that all mandatory information for a soft drink for the (bi-lingual) Belgian market presented in the proposed font size, would cover an entire tin can in text hardly leaving any space to present the brand name.

courageous. The burden the proposal is likely to create is perceived as too high in comparison to the advantages. The Commission should go for full codification or include the most necessary changes in the existing General labelling directive 2000/13/EC.

8. Supportive schemes

Can the legislator create supportive schemes for food businesses in general and SMEs particular?

8.1 Problem statement

As elaborated in the Specifications Call for Tenders[136] the European Commission considers development of an 'EU Food product origin marking scheme' as a mark of distinction.

The number of conceivable options and combinations of origin labelling is almost endless. Diagram 6 indicates 32 of them. Half of these are of a public law nature, half of a private law nature. Six options are included in the terms of reference, all of them of a public law nature. At least two options exist in current EU legislation, a third is in preparation (not for food). Some options have been dealt with by the ECJ. For most products the current situation is that no regulation exists and that therefore (food) business operators are free within the general legal limits like the ban on misleading information to choose to indicate an origin. In addition to the terms of reference the Commission expressed the desire to be informed about the option to indicate the SME producer on private labels.

The envisaged marking scheme is intended to create a 'mark of distinction'. This means that the message is to be conveyed that the fact that a product comes from the EU (or other area as indicated in Diagram 6) ensures purchasers of its quality.

8.2 Definition of terms

The term 'label(ling)' has different meanings in an economic or a legal context.[137] In this report the word label is used in the narrow legal meaning following from the definition proposed in the Commission Proposal for a Regulation of the European Parliament and of the Council on the provision of food information to consumers,[138] Article 2 (2)(1):

> *"label" means any tag, brand, mark, pictorial or other descriptive matter written, printed, stencilled, marked, embossed or impressed on, or attached to, a container of food.'*

In simple words: label is the actual piece of paper glued to the package in which the food product is being sold. In this report labelling means: indicating information on the said piece of paper.

---

[136] No ENTR/2007/020.
[137] See: Poppe *et al.*, 2009.
[138] COM(2008) 40 final, discussed above.

Diagram 6. Options for origin marking.

	Producer	Regional	MS	EU	3rd	MS + 3rd	EU + 3rd	EU + MS + 3rd
Public								
Mandatory								
op. x				op. 2-2	op. 2-1 COM (2005) 661	beef, FFV 8-74; C-207/83	op. 3	
Voluntary								
op. x		PDO PGI C-6/02	C-325/00	op. 1-1/ 2-1	op. 2-2		op. 1-2	
Private								
Self regulation								
No regulation								
intel inside		current	current	current	current	current	current	current

MS = Member State; 3rd = third country.
op. = option in terms of reference; op. x = additional option; C- = case law; FFV is fresh fruit and vegetables.

8.3 Theoretical background

Literature shows two different perspectives on establishing and communicating the origin of products. One perspective is on imported products, the other on domestic production. The perspective on imported goods can be labelled as 'rules of origin'.[139] The perspective on domestic production can be labelled: 'origin marking'.[140] Together they will be called hereafter 'origin requirements' (ORs).

---

[139] Lang and Gaisford, 2007.
[140] DG Trade, 2006b.

8.3.1 Rules of origin

'Rules of Origin' (hereafter: ROOs) identify the country where a good is deemed to have originated. One of the primary roles of ROOs is to prevent tariff circumvention, or to put it more generally; to reserve preferential treatment to goods originating in the country with which such treatment has been agreed upon and to prevent goods from other countries from taking advantage as well.

Preferential and non-preferential ROOs
ROOs are distinguished into preferential and non-preferential ROOs.[141] National governments implement non-preferential ROOs unilaterally for keeping trade statistics, country of origin labelling, government procurement, anti-dumping and the like. Preferential ROOs define the products that are eligible for preferential (e.g. tariff free) access between member countries (MC) of a free trade area (FTA) or similar preferential trading agreement.

Criteria
ROOs generally apply one or more of three types of criteria to determine the origin of a product:
1. change of tariff classification
    The Harmonized Tariff System[142] separates products into major categories called chapters and then further separates them into subcategories called classifications, headings and subheadings. For a product to be considered to originate in a given country, it must go through a change in category as specified in advance, due to a production process in the said country.
2. *ad valorum* or percentage test
    This test calculates the percentage of the product's costs by adding up the country specific inputs and comparing them with the total cost of production. To be accepted as originating in a certain country, the percentage attributable to this country has to meet a designated level. Often this calculation employs roll-up and roll-down provisions and a de minimis rule.
3. specified operations test
    This test requires the product to have undergone specific steps in manufacturing or processing in a certain country to originate in that country.

WTO Agreement on Rules of Origin
The applicable WTO context is Article IX of the GATT in connection with the WTO Agreement on Rules of Origin (hereafter: AROO).[143] AROO aims at long-term harmonisation of rules of origin, other than rules of origin relating to the

---

[141] Lang and Gaisford, 2007.
[142] In the World Customs Organization's (WCO) Convention on the Harmonized Commodity Description and Coding System.
[143] Information from www.wto.org.

granting of tariff preferences, and ensuring that such rules do not themselves create unnecessary obstacles to trade.

AROO set up a harmonisation programme, based upon a set of principles, including making rules of origin objective, understandable and predictable.[144] Until the completion of the harmonisation programme, contracting parties are expected to ensure that their rules of origin are transparent; that they do not have restricting, distorting or disruptive effects on international trade; that they are administered in a consistent, uniform, impartial and reasonable manner, and that they are based on a positive standard (in other words, they should state what does confer origin rather than what does not).

An annex to AROO sets out a 'common declaration' with respect to the operation of rules of origin on goods which qualify for preferential treatment. If rules of origin comply with AROO, they are acceptable under WTO-law.

Rules of origin are compatible with WTO-law, within the limits set by AROO.

8.3.2 Origin marking

General
A *mark of origin* ('made-in' marking) is a permanent sign (e.g. etching, moulding) on a product which signals its geographical origin. Origin marking (OM) may be required for imports and/or domestic goods, it may cover certain sectors or all goods, or legislation may just set a framework for the voluntary use of OM. OM is distinct from labelling requirements.

The *purpose* of OM is transparency and informed purchase decisions by consumers, and to reduce the incidence of fraudulent or misleading indications that would undermine the reputation of producers. From this spring certain practical features. The marking must reach the final purchaser. It must be sufficiently permanent and indelible. It must be visible, clear, and easily understood by the ultimate purchasers.[145]

Country of origin labelling
'Country of Origin Labelling' (hereafter: COOL) seems to be the expression used for origin marking regarding food products. It refers to the requirement for retailers to inform consumers of the country of origin of a product at the point of sale. The

---

[144] The work is conducted by a Committee on Rules of Origin (CRO) in the WTO and a technical committee (TCRO) under the auspices of the Customs Cooperation Council in Brussels.
[145] DG Trade 2006a.

argument commonly given in favour of COOL is that consumers prefer domestic products to imported or that consumers in export-destination countries consider the origin an asset.

Opponents argue that consumers seem to have little interest in COOL – otherwise it would have developed spontaneously. COOL increases costs of labelling, record keeping and procedures necessary to support the requirements, and thus decreases food choice.[146]

Previous LEI research on COOL in the meat sector shows that Dutch consumers hardly differentiate between meat products originating in the Netherlands or other western European countries. However they consider these superior to meat products from eastern European or third countries.[147] For this reason processors chose the latter products for the types of preparation that lift the obligation to label the origin of beef. In general in the Netherlands COOL requirements have not had an influence on consumer demand. German and French consumers however exhibit a preference for national products.

A new standard in the Australia New Zealand Food Standards Code requires packaged food to carry a separate statement identifying the country where the food was produced, made or packaged. The new country of origin labelling standard for packaged food has been phased in over a two-year period from 8 December 2005.

The USA 2002 Farm Bill requires country of origin labelling for beef, lamb, pork, fish, perishable agricultural commodities, and peanuts. The entry into force has – except for fish and shellfish – been delayed several times. COOL has become effective on 16 March 2009.

8.4 Current and coming EU ORs

8.4.1 General

Community Customs Code
The Community Customs Code[148] defines the non preferential origin of goods for the purpose of applying the Customs Tariff of the EC; applying other measures relating to trade in goods and the preparation and issue of certificates of origin[149] and preferential origin.[150]

---

[146] Krissoff *et al.*, 2004.
[147] Van Horne *et al.*, 2006.
[148] Regulation 2913/92 of 12 October.
[149] Articles 22-26.
[150] Article 27.

Goods originating in a country are:
- goods wholly obtained (including harvested) or produced in a country or on board a factory ship;
- goods that underwent their last, substantial, economically justified processing or working resulting in a new product or representing an important stage of manufacture.

1980 proposal
A proposal for a directive on the approximation of the laws of the member states relating to the indication of the origin of certain textile and clothing products, which the Commission submitted to the Council in 1980,[151] received a negative opinion from the Economic and Social Committee in 1981.[152] Although the Committee believed that it was essential for consumers to be able to make their buying decisions in the light of adequate information, it considered that the indication of a product's country of origin did not fill a genuine consumer need; other information, such as price, composition, grade, quality and instructions for use, were more important. The Commission agreed and withdrew the proposal.

Proposed EU origin marking scheme
In a series of working documents[153] DG Trade proposed the introduction of a 'made-in' marking scheme. According to DG Trade such a scheme may cover two types of goods: (a) imported goods, and (b) domestic production for the internal market.[154] Origin marking can be regulated in either a voluntary or a compulsory manner.

The main objectives of such a scheme could be:
- To introduce greater homogeneity and clarity across the EU internal market through an EU-wide instrument.
- To provide comprehensive and accurate information to consumers on the country of origin of products, and in a way that would promote in parallel the image and attractiveness of EU products, e.g. as possessing a certain quality, reliability or style. This could in turn contribute to enhance European companies' competitiveness in the global arena against foreign competitors producing at lower costs and investing less in consumer protection.
- To reinforce the concept of 'made in the EU' as a means to consolidate the image and recognition of the EU's single customs union and single market, both for its own sake and as an element in the broader efforts to gain greater recognition of the customs union in the international framework.

---

[151] Official Journal 1980 C 294, p. 3.
[152] Official Journal 1981 C 185, p. 32.
[153] Without reference number. See the reference list for DG Trade 2003, DG Trade 2004, DG Trade 2005, DG Trade 2006a and DG Trade 2006b.
[154] DG Trade 2003.

- To provide an additional tool to combat consumer deception and use of fraudulent origin marking considering the current systematic practices of false labelling and fraud.

Proposal Regulation on indication of country of origin on imported products
After consultation of stakeholders, the Commission submitted a proposal[155] for a 'Council Regulation on the indication of the country of origin (CO) of certain products imported from third countries'. This draft Regulation aims at the introduction of a compulsory origin marking scheme covering a number of sectors which see benefit in the initiative, and applicable to *imported goods* only. This is the option which, according to the Commission, on balance, takes best into account the interests of the larger share of stakeholders (industry, trade unions, and part of the consumer movement), it is an option which limits any costs and negative effects for other interested parties (EU industries that have delocalised their production, traders), and which ensures at the same time a positive impact as regards the policy objectives of the initiative.

The proposed regulation requires the country of origin to be marked on goods using the words 'made-in' together with the name of the country of origin. The country of origin is determined on the basis of the criteria set out in the Community Customs Code for non preferential origin. The regulation is intended to apply to industrial products listed in an annex such as leather, footwear, textiles, ceramics, jewellery, furniture and brooms. It explicitly excludes foodstuffs from its scope.

Proposed non-EU origin marking does not apply to food.

8.4.2 Food

General labelling requirements
It is a general principle of food law, that labelling may not mislead the consumer, see: Articles 8 and 16 of the General Food Law. This principle has been applied to the origin of foodstuffs in the General labelling directive.

The General labelling Directive 2000/13 (as amended) requires in Article 3(1)(8) particulars of the place of origin or provenance to be labelled where failure to give such particulars might mislead the consumer to a material degree as to the true origin or provenance of the foodstuff. This requirement elaborates on Article 2(1)(a)(i) stating that the labelling and the methods used must not be such as could mislead the purchaser to a material degree particularly as to (as far as relevant in the current context) the origin or provenance of the foodstuff.

---

[155] COM(2005) 661 final.

Food labelling may not mislead the consumer as to the origin of the product.

Beef and veal
In addition to the general labelling requirements, Regulation 1760/2000 includes some compulsory provisions and some voluntary measures for the labelling of beef and veal.[156] The regulation, replacing earlier legislation, aims at maintaining and strengthening consumer confidence in beef and avoiding misleading them by ensuring that information is made available by sufficient and clear labelling.[157]

Some of the compulsory requirements are relevant in the context at issue. Operators and organisations marketing beef in the Community are obliged to label it in accordance with Article 13 of Regulation 1760/2000. The compulsory labelling system must ensure a link between, on the one hand, the identification of the carcass, quarter or pieces of meat and, on the other hand, the individual animal or, where this is sufficient to enable the accuracy of the information on the label to be checked, the group of animals concerned.

Further, the label must show:
• in which country the animals were born;
• in which country the animals were fattened/bred;
• a code for the country and the slaughterhouse where the slaughter took place.

These requirements are very similar to the specified operations test mentioned above,[158] except that they can lead to more than one country of origin.

For minced meat some derogations apply.[159] For example, the three issues mentioned above may be replaced by an indication of the country (Member State or third country) where the meat was prepared and 'origin' if this is not the same country. Beef imported from a third country may be labelled 'origin: non-EC' and 'slaughtered in: (name of third country)' if all the required information is not available.

For beef and veal the countries must be labelled where the animal was born, bred and slaughtered.

---

[156] Council Regulation 1760/2000 17.7 2000, O.J. (L 204), 1-10.
[157] Recital 4. See also: Van der Meulen and Freriks, 2007.
[158] See section 8.3.1.
[159] Article K1.

This regulation has been issued as part of the responses to the BSE-crisis. This background seems to justify the assumption that labelling the country where certain steps in the production chain have taken place, is meant to dissociate the animals concerned from areas stricken by BSE. At the time of enactment of the regulation, that was mostly the UK. Beef and veal labelled 'origin UK' was difficult to sell.

Evaluations show as a result of this legislation (and its predecessor) a tendency towards localisation of slaughter, meaning a decrease of export of cattle and an increase of slaughter in the country where the animal was bred.[160] If Member State of origin labelling where to be substituted by EU-labelling, this would mean a considerable reduction of burden on business.[161]

As indicated in section 8.3.2, it is also believed that beef originating from a less popular country is preferably used in meat products for which an exception to the labelling requirement exists.

Fresh fruit and vegetables
Regulation 1182/2007 laying down specific rules as regards the fruit and vegetable sector states that certain products[162] – fruits, vegetables and nuts – which are intended to be sold fresh to the consumer, may only be marketed if they are sound, fair and of marketable quality *and if the country of origin is indicated*. Regulation 1234/2007 provides marketing standards for other agricultural products to relate among other things to origin and labelling.[163]

The only explanation on the background of the required country of origin labelling, is found in the recitals of Regulation 2200/96: 'whereas in particular, consumer requirements as regards the characteristics of fruit and vegetables mean that the origin of products should be included in the labelling up to and including the final retail stage.'[164]

---

[160] European Commission 2004; Van Horne *et al.*, 2006.

[161] Van Horne *et al.*, 2006.

[162] The products listed in Article 1(2) of Regulation 2200/96: tomatoes, onions, shallots, garlic, leeks and other alliaceous vegetables, cabbages, cauliflowers, kohlrabi, kale and similar edible brassicas, lettuce (*Lactuca sativa*) and chicory (*Cichorium* spp.), carrots, turnips, salad beetroot, salsify, celeriac, radishes and similar edible roots, cucumbers and gherkins, leguminous vegetables, shelled or unshelled, other vegetables, other nuts, whether or not shelled or peeled, excluding areca (or betel) and cola nuts, plantains, figs, pineapples, avocados, guavas, mangos and mangosteens, citrus fruit, table grapes, melons (including watermelons) and pawpaws (papayas), apples, pears and quinces, apricots, cherries, peaches (including nectarines), plums and sloes, other fruit, saffron, thyme, basil, melissa, mint, origanum vulgare (oregano/wild marjoram), rosemary, sage, locust (or carob) beans. The sequence of the listing follows from the CN codes not included in this footnote. Some elaborations and exceptions present in the text of the law have also been omitted from this footnote.

[163] Article 113(2)(b) Regulation 1234/2007.

[164] From recital 5 of Regulation 2200/96.

For fresh fruits and vegetables the country of origin must be labelled.

PDO/PGI

Regulation 510/2006 on the protection of geographical indications and designations of origin for agricultural products and foodstuffs, establishes rules for the protection of certain designations of origin (PDO) and geographical indications (PGI) of agricultural and food products. The regulation provides opportunities for small-scale producers to use these quality symbols as a means of promoting their products, without the long and costly process of obtaining a trademark for their product.[165]

A PDO, 'protected designation of origin' (e.g. 'Parma ham' or 'Camembert de Normandie'), includes the name of a region or specific place that is used to describe an agricultural product or foodstuff originating in that region or place. The PDO should designate only those products that exhibit the quality or characteristics, which are essentially, or exclusively due to a particular geographical environment with its inherent natural and human factors, and the production, processing and preparation which take place in the defined geographical area.

A PGI, 'protected geographical indication' (e.g. 'Ardennen ham' or Danablu cheese), includes the name of a region or a specific place used to describe an agricultural product or foodstuff originating in that region or place which possesses a specific quality, reputation or other characteristics attributable to that geographical origin and the production and/or processing and/or preparation which take place in the defined geographical area.

Protected designations communicate qualities linked to geographical origin.

To obtain a 'protected designation of origin' the area must be precisely defined, all stages of production, processing and preparation, from the raw materials to the finished product, must take place in the area that lends its name to the product. The characteristics of the product must be essentially or exclusively due to the place of origin. The ties between area and product are much less strict for a PGI. The product must have been produced in the indicated geographical area. However, it is sufficient that only one of the stages of production has taken place in that area. Also, the characteristics of the product do not have to be essentially or exclusively due to the indicated area. For a PGI it is sufficient that only one characteristic of the product can be attributed to the area, for instance its reputation.

---

[165] Van der Meulen and Van der Velde, 2005.

Nevertheless both types of designation clearly communicate quality attributes to the consumer.[166]

8.4.3 EU Labelling Proposal

The EU Labelling Proposal[167] incorporates the definition of country of origin from the Community Customs Code as described in section 8.4.1. The proposal takes as a general principle that food information shall not be misleading as to the country of origin or place of provenance.[168] As a consequence, indication of the country of origin or place of provenance is mandatory where failure to indicate this might mislead the consumer to a material degree as to the true country of origin or place of provenance of the food, in particular if the information accompanying the food or the label as a whole would otherwise imply that the food has a different country of origin or place of provenance.[169]

The labelling proposal clarifies the concept of origin.

The Chapter on voluntary food information explicitly deals with origin labelling in the situations where it is not mandatory:

The paragraphs 2, 3 and 4 of Article 35 read:

'2. Without prejudice to labelling in accordance with specific Community legislation, paragraphs 3 and 4 shall apply where the country of origin or the place of provenance of a food is voluntarily indicated to inform consumers that a food originates or comes from the European Community or a given country or place.

3. Where the country of origin or the place of provenance of the food is not the same as the one of its primary ingredient(s), the country of origin or place of provenance of those ingredient(s) shall also be given.

4. For meat, other than beef and veal, the indication on the country of origin or place of provenance may be given as a single place only where animals have been born, reared and slaughtered in the same country or place. In other cases information on each of the different places of birth, rearing and slaughter shall be given.'

8.5 EU case law on national ORs

Some of the options indicated in Diagram 6 appeared in legislation of Member States and were judged by the European Court of Justice (ECJ) for conformity

---

[166] See in great detail on indications of origin: O'Connor 2004.
[167] Article 2(3) COM(2008) 40 final.
[168] Article 7(1)(a).
[169] Article 9(1)(i).

with the ban on measures having an equivalent effect to quantitative barriers to trade (now[170] Article 28 EC Treaty).

The 1974 Dassonville case[171] deals with a requirement of a certification of origin to be issued by the authorities in the exporting state. This requirement is judged by the ECJ to constitute a measure of similar effect as a quantitative barrier to trade:

> *'In the absence of a community system guaranteeing for consumers the authenticity of a product's designation of origin, if a Member State takes measures to prevent unfair practices in this connexion, it is however subject to the condition that these measures should be reasonable and that the means of proof required should not act as a hindrance to trade between Member States and should, in consequence, be accessible to all community nationals.'*

A Court ruling from 1985[172] deals with national legislation prohibiting the retail sale of certain products imported from other Member States, unless they bear or are accompanied by an indication of origin. The Court is highly critical. For several reasons it concludes that the measure at issue is a barrier to trade for which no justification can be found in EC law. The Court argues that such a requirement would necessarily increase the production costs of the imported article and make it more expensive. Further it has to be recognised that the purpose of indications of origin or origin marking is to enable consumers to distinguish between domestic and imported products and that this enables them to assert any prejudices which they may have against foreign products. In short, the provisions in question are liable to have the effect of increasing the production costs of imported goods and making it more difficult to sell them. The court also notes that if the national origin of goods brings certain qualities to the minds of consumers, it is in manufacturers' interests to indicate it themselves on the goods or on their packaging and it is not necessary to compel them to do so. In that case, the protection of consumers is sufficiently guaranteed by rules which enable the use of false indications of origin to be prohibited. Such rules are not called into question by the EEC treaty.

In 2002 the Court ruled that[173] by awarding the quality label 'Markenqualität aus deutschen Landen' (quality label for produce made in Germany) to finished products of a certain quality produced in Germany, the Federal Republic of Germany has failed to fulfil its obligations under (now) Article 28 of the EC Treaty.

The contested scheme has, at least potentially, restrictive effects on the free movement of goods between Member States. Such a scheme, set up in order to promote the distribution of agricultural and food products made in Germany and

---

[170] Previously Article 30.
[171] ECJ 11 July 1974, Procureur du Roi v. Dassonville, case 8-74.
[172] ECJ 25 April 1985, Commission v. UK, case 207/83.
[173] ECJ 5 November 2002, Commission v. Germany, case C-325/00, ECR 2002, p. I-09977.

for which the advertising message underlines the German origin of the relevant products, may encourage consumers to buy the products with the quality label to the exclusion of imported products.

The fact that the use of that quality label is optional does not mean that it ceases to be an unjustified obstaclc to trade if the use of that designation promotes or is likely to promote the marketing of the product concerned as compared with products which do not benefit from its use.

In 2003 the Court ruled that protected designations of origin may not be introduced by national legislation but may only be afforded within the framework of Regulation 2081/92 (now Regulation 510/2006).[174]

The ECJ considers origin marking a trade barrier.

The Court consistently rules both with regard to mandatory and with regard to voluntary legislative schemes on the protection of indications of origin, that these constitute barriers to trade infringing on Article 28 EC Treaty. While Article 28 EC Treaty[175] seems to be more strict than Article VIII GATT, the position of the EU to argue that the envisaged marking scheme is compatible with Article VIII GATT, is not greatly improved by the case law stating in no uncertain terms that they constitute trade barriers.

8.6 Co-labelling

8.6.1 Concept

The idea of indicating the identity of the (SME) producer on the label of a product brought to the market under the name of another business; the (private) label holder, can be christened: chain transparency or co-labelling. It gives the consumer some information regarding the composition of the food production chain that led up to the product as offered for sale.

In current EU food legislation, some examples exist of legislation requiring the labelling of information relating to the make-up of the chain the food in question has gone through.

---

[174] ECJ 6 March 2003, Commission v. Germany, Case C-6/02.
[175] Available at: http://eur-lex.europa.eu/LexUriServ/LexUriServ.do?uri=OJ:C:2006:321E:0001:0331:EN:pdf.

8.6.2 Elements in current legislation

Health marking
On the basis of Regulations 853/2004 and 854/2004 meat has to bear a health mark indicating the slaughterhouse where the animal has been slaughtered.

Traceability
EU food law encompasses a mandatory general traceability system.[176] This system requires traceability information to be available one step up and one step down only. Chain transparency or COOL would require an intact paper trail with regard to the aspects chosen as criteria.

GMO traceability
In contrast to the general traceability requirements, products originating from genetically modified organisms must be accompanied by an intact paper trail from farm to fork.[177]

8.6.3 WOK

Since 2002 the Netherlands' consumers' organisation *Consumentenbond* has been lobbying for introduction of a chain transparency act (Wet Openbaarheid Ketens: WOK). This idea seems to be gaining support from politicians and consumer NGOs in other countries. The act as proposed by the Consumentenbond would give consumers the right to demand from businesses information on social aspects of their product. Social aspects often relate to the history of the product in the chain.

8.6.4 Co-labelling

The notion of chain transparency is not entirely new. Apart from SMEs consumers have also voiced wishes in this direction. Compliance with certain existing requirements may help to have relevant information available for co-labelling.

8.7 An additional protected designation?

The instrument the Commission is looking for, is meant to support the market position of SMEs in the food sector in general. Experience shows that the protected designations (PDO, PGI) provide goods with a quality reputation more easily attainable for SMEs than branding on the basis of a trade mark. These designations are however only attainable for specific region-product combinations. The vast majority of SMEs will not be able to meet these.

---

[176] Article 18 Regulation 178/2002.
[177] Regulation 1830/2003. On this issue see: Van der Meulen, 2008.

Is there an aspect that consumers may appreciate in all SME products? The only common aspect that comes to mind is the scale of production. Indeed groups of consumers are known to appreciate small scale: the human size. Maybe it would be worthwhile to initiate empirical research into the question if there is a market for a protected designation of guaranteed small scale production.

The counterargument would be the same as was voiced with regard to COOL. Businesses are free to include such information in the label. If they would expect an advantage of such labelling, it would already be present in the marketplace.

8.8 Benefits and risks

8.8.1 Benefits

The most important benefit usually expected from any form of origin marking is the attention from consumers who consider a certain origin an asset. Businesses exploit this by investing in their brand reputation. A product originating from the business concerned enjoys reputation. Protected designations of origin do the same on a collective basis. In so far as empirical data are available geographical origin (without the additional quality guarantee of a PDO/PGI) mainly appeals to consumers for chauvinistic reasons related to a Member State, not to the EU. This benefit applies to the home country only. Businesses are free to mention the state of origin of their product. The survey we conducted in this second food competitiveness study shows that businesses value the mentioning of their name on the label of the product (Diagram 7). Producers seem to believe that they can build a certain reputation if they are mentioned as the producer on the label of the brand holder.

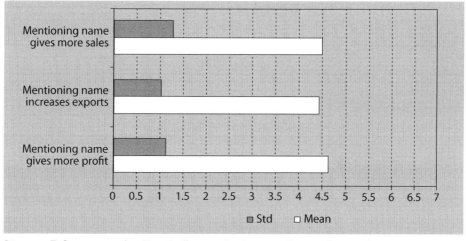

Diagram 7. Scores on a 1 – 7 scale (1 = totally disagree; 7 = totally agree).

8.8.2 Risks

Transaction costs
Compliance with ORs entails significant transaction costs for many businesses, which reduce both trade volumes and gains from trade.[178]

Protectionism
ORs have the potential to act as instruments for protectionism in two related ways.[179] First stricter rules of origin typically reduce the extent of new trade creation in final goods between member countries within a free trade area. Second, stricter rules of origin also lead to greater trade diversion with respect to trade with outside countries in intermediate goods because there are increased incentives to produce intermediate goods within an FTA so that final goods will meet the preferred ORs.

Boycott
Political disagreement sometimes translates into a consumer boycott of products or authorities' restrictions on imports. These may have lasting effects. In the mid 1980s, for example, in the Netherlands high quality in wine was strongly associated with origin in France. After French government agents (DGSE) bombed the Greenpeace ship protesting nuclear tests at Moruroa killing a crew member of Dutch/Portuguese nationality on 10 July 1985 a consumer boycott of French products followed. As a collateral, Dutch consumers 'discovered' non-French wines that to the detriment of France acquired a market share they still hold today.

In our previous round of interviews, a Dutch dairy company related that after (EU) rejection in Rotterdam harbour of shrimps contaminated with chloramphenicol, some Asian inspection agencies claimed to find contaminations in Dutch dairy products. The business resorted to supplying its costumers from its German branch, thus circumventing the suspicion of Dutch products.

As indicated above the mandatory country of origin labelling of beef may in practice be perceived as non-UK-labelling. These examples show that COOL holds the potential to backfire.[180] Experience with traceability systems shows that the bigger the lot, the bigger the damage in case of a recall. Similarly the introduction of an EU-marking scheme may increase the size of the lot affected by negative appreciation of a single Member State's actions.

---

[178] Lang and Gaisford, 2007.
[179] Lang and Gaisford, 2007.
[180] A more recent example, finally, is the boycott of Danish products by Islamic consumers after a Danish artist published cartoons considered offensive to Islam.

Co-labelling

Co-labelling can be perceived from two different perspectives. The perspective of businesses producing their own brands and producing for private label holders as well and the perspective of businesses producing only for another business' label. The former may not be overly pleased by chain transparency as the private label may be perceived as undermining their own label: premium brand's quality at private label price. Given the choice in a voluntary scheme, they will probably choose not to be mentioned. Businesses on the other hand that depend on a private label will not be in a bargaining position to exercise their rights under a voluntary scheme as the private label holder is likely to prefer to do business with operators not invoking their rights.

In the survey of the second food competitiveness study, we presented three options to stakeholders (Diagram 8):
- a mandatory system requiring the name(s) of the processor(s) to appear on the label of the end product;
- a voluntary system giving processors the right to demand mention of their name on the label;
- a voluntary system giving the end-producer the choice to print names of processors on the label.

None of these models was greeted with much enthusiasm.

Liability

Food businesses can be held liable if they bring unsafe products to the market. Discussions surrounding the introduction of traceability show that some fear existed that this would lead to increased liability risks. There are two sides to this issue. For businesses upstream in the food production chain, indeed this risk will increase. For businesses downstream, the situation is the inverse. If they can indicate the business that caused the problem, it is easier for them to pass on liability. Under product liability law,[181] the retailer can pass on liability to the producer by informing the consumer of the producer's identity. This is

Diagram 8. Opinions on co-labelling (1 = totally disagree; 7 = totally agree).

		Mandatory	Voluntary for processor	Voluntary for end-producer
N	valid	28	28	29
Mean		2.89	3.25	3.03
Std. deviation		2.114	2.255	1.936

---

[181] See: Directive 85/374. On liability in the chain see also COM(2007) 469.

not possible for the private label holder. The business that has its name on the label *is* the producer within the meaning of product liability law. If, however, the manufacturer is separately mentioned on the label it is conceivable that the consumer will choose to address the manufacturer instead of brand holder or will accept being redirected by the brand holder to the producer. We cannot put it beyond the average consumer to believe if so informed by the retailer that in product liability law the liable 'producer' is the one who produced the product. It takes some understanding of the law to realise that in the context of product liability 'producer' is a wider concept than in common speech.[182]

8.9 Concluding remarks

From a legal point of view all options are possible in EU law: EU-labelling, country of origin labelling and business of origin labelling (co-labelling). The WTO Agreement on Rules of Origin provides sufficient options.

The aim of the scheme is to support businesses. Businesses that expect an advantage from EU or Member State origin labelling are free to do so. The only protection needed is to ensure that origin labelling has a well defined meaning. The food labelling proposal takes care of this issue by referring to the Community Customs Code.

For co-labelling matters stand differently. The intended beneficiary of the scheme is not the business doing the labelling, but a business earlier in the chain. Such a scheme can only be expected to be effective if it is mandatory. The limited data available at this point do not show much support for such a scheme.

---

[182] See Article 3(1) of Directive 85/374.

9. Good administrative practices

What can be done by administrative authorities to alleviate the burden from EU food law? Probably the most striking addition of the outcomes of the current research to those of the first food competitiveness project, is the realisation that part of the burden felt by food businesses from the food regulatory framework does not result from the makeup of the framework as such, but from shortcomings in compliance by the authorities applying the framework.

Neither the Commission nor the Member States' experts meticulously respect the legal limits set to their powers. In part this remains hidden behind reluctant application of the principles of transparency and the duty to give reasons.[183] Constraints on authorities are such that deadline discipline does not get the chance to develop into a way of life.

Improved compliance by public authorities would alleviate the burden on businesses from EU food law.

We have also seen that authorities are a very important – but not always accessible – source for businesses of understanding food law. If both EU and national authorities realise this and make an effort to take responsibility through compliance assistance, this may be of considerable help for SMEs and big businesses in the EU food sector.

Administrative practices can contribute in two ways to reducing the burden on food businesses, through improved compliance with the law by the authorities and through compliance assistance to help businesses deal with their obligations.

---

[183] Article 253 EC Treaty.

10. Conclusions

The restructuring of EU food law as envisaged in the White Paper on Food Safety[184] is nearing its completion. The time seems ripe for fine tuning and rethinking on the basis of experience gathered so far and also taking account of the findings in this study. Compliance emerges as a key issue. Compliance by authorities with the legal requirements and compliance by the legislator with the EC Treaty and general principles of food law.

The objective of this study was to identify food legislation that:
• impedes the placing of a food product on the market with the ensuing consequences for competitiveness;
• raises unjustifiable or unnecessary costs to economic operators, which lead *ipso facto* to a price increase of the end food product;
• prolongs delays prior to the placing on the market of end food products causing detrimental effects on competition, whereas this legislation does not bring substantial benefits for public health, consumer protection or the environment.

10.1 Food law and competitiveness

As competitiveness is outside the scope foreseen for food legislation in the EU, there is a serious risk that EU food sector competitiveness is hampered by EU food law. Compliance with Article 157(1) of the EC Treaty would alleviate this problem. Legislation should provide the necessary framework for doing so.

10.2 Premarket approval schemes

Premarket approval schemes are an exception to the principle of proportionality. This principle requires scientific proven necessity for measures forming barriers to business. Premarket approval schemes reverse the burden of proof regarding food safety. The principle of proportionality, the SPS Agreement and Article 6 GFL lay this burden on public authorities, approval schemes shift it to food businesses. This leads *ipso facto* to a price increase of the end food product. According to the European Commission (in its Communication on the application of the precautionary principle) premarket approval schemes are based on the precautionary principle.

Case law consistently holds, as does Article 7 GFL, that precautionary measures may only be taken and maintained as long as conclusive risk assessment cannot be made. In situations where conclusive risk assessment is available, no precautionary measures can be taken. Exclusive market approval, second stage market approval procedure after initial positive risk assessment at member state level, market

---

[184] European Commission 2000a

approval procedure after positive risk assessment by a trustworthy non-EU risk assessment authority (like JECFA) are situations where the availability of conclusive risk assessment should be taken into account.[185] Legislation should provide the necessary framework for doing so.

10.3 Food safety objectives

The findings in this report on food safety objectives are based on desk research and the opinion of one single interviewee (supported by some of his colleagues). This single opinion has been convincing for the authors of this report.

The critical issue in food safety objectives is zero-tolerance levels. Measuring capabilities have developed to such an extent that in any sample of sufficient size almost anything can be detected at levels increasingly close to zero.

Decisions to destroy food can exclusively be justified on the basis of scientific risk assessment. Zero-tolerance levels, however, are not always the result of scientific risk assessment. Often the safety objective is a combination of attainable levels under good practice as established in an approval procedure on the one hand and risk assessment on the other hand. A zero-level may result from insufficient scientific support of the product by the applicant for approval or from a negative outcome of this procedure. In both situations enforcement of the zero-level is a sanction, not a science-based risk management measure. Such a sanction can hardly be accepted as ethical if it results in the destruction of food. Such a sanction is also indiscriminate in that it does not target the business that may have infringed by using unapproved products,[186] but the business last handling the product.

10.4 Food hygiene

If the zero-level *is* the result of risk assessment in the sense that a safety level could not be established, it is by definition precautionary. Under Article 7 GFL such level may only be provisional and further risk assessment is required (without undue delay). If zero-tolerance levels are not or no longer scientifically justified, they place an unreasonable burden on food businesses.

---

[185] Outside the food sector some examples exist of international Mutual Recognition Agreements regarding the acceptance or recognition by contracting parties of other parties' conformity assessments. See for instance the MRA EU-USA. The argument above, however, is based on the principles of EU food law. Under these principles it is not necessary – desirable but not required – that recognition be mutual.

[186] There is also the generally applicable problem, mentioned previously, that adventitious environmental contamination, whether by naturally occurring compounds or unwitting and unknown exposure to man-derived components, can render illegal through zero-tolerance what is reasonably thought to be acceptable, well-produced food.

At closer inspection food hygiene legislation seems to hold solutions to the problems perceived by small scale traditional producers. The problems are not in the law but in the limited knowledge of the law by businesses and authorities alike. However, for financial reasons, traditional methods may be difficult to uphold if they are considered of higher risk than industrial methods and traditional producers are charged for high frequencies of controls.

10.5 Food labelling

The improvements in substance of food labelling law seem too limited to warrant the large scale change in structure of the legislation that has been proposed.

10.6 Country of origin labelling

WTO law does not seem to be an obstacle as such for the introduction of a system of COOL in the EU. Given the position of the European Court of Justice that COOL in virtually all its occurrences constitutes a barrier to trade, some explaining will nonetheless have to be done.

It is remarkable that businesses are little inclined to spontaneous COOL. Apparently they do not expect great advantages. The limited empirical data available so far indicate that for some consumers, Member State origin has greater significance than EU origin. It would however go against the grain of the internal market to require Member State origin to be labelled.

10.7 Co-labelling

Requirements on chain transparency through co-labelling will impose additional administrative burdens on businesses. A voluntary scheme will not work for the intended audience as in most cases they will lack the bargaining power to profit from it. A mandatory scheme will not be appreciated by businesses fearing to undercut their premium brand by showing their name on a private label. They may however solve this problem by using a different business name when producing for a private label. This study has not produced sufficient support to recommend co-labelling to be introduced.

11. Recommendations

11.1 For the legislature

One of the specific objectives of this part of the study is to identify existing pieces of legislation in the food industry sub-sector aimed at improving public health, consumer protection or environment that obviously would not have been adopted or would have been worded differently if genuine impact assessment had been carried out prior to their adoption at Community level. The aim is to indicate opportunities for the EU legislators and executive to remove avoidable obstacles for the food industry as a means to reduce regulatory burdens and/or enhance competitiveness. The basic finding of this study is that improved (understanding and) compliance by authorities with the EC Treaty and the principles of food law, will benefit businesses.

- Article 5 GFL setting the objectives of EU food law should be adjusted to mirror Article 157 EC Treaty.
- Efforts to harmonise, codify and simplify food legislation should continue.
- A regulatory rhythm should be introduced meaning that, unless urgency dictates otherwise, new food legislation will enter into force together on only one date in any year.
- Deadlines for authorities should be fatal.
- The proliferation of premarket approval schemes should be reversed. Respecting the precautionary principle as set out in Article 7 GFL, such schemes should be limited to foods regarding which food safety concerns are proven to be more than hypothetical.
- Respecting Article 6 GFL, such legislation should be preceded by advice from EFSA.
- Respecting the precautionary principle as set out in Article 7 GFL, businesses should be spared the burden to prove the safety of a food in situations where conclusive risk assessment already exists, more in particular if it is already in the hands of the EU authorities.
- To this end legislation should introduce recognition of risk assessment by competent bodies.
- The proposed new regulations on premarket approval schemes should be notified (and justified) under the WTO SPS Agreement.
- A fast track procedure should be introduced to decide on the applicable procedure.
- Political decisions on premarket approval should be laid down in legislation of a general nature. Application of this legislation to specific food should be a task of an administrative nature without comitology, entrusted to an administrative body accountable to the European Parliament but without participation of the Parliament in the procedure.
- Complying with Article 6 GFL, zero-tolerances should on the basis of sound risk assessment be reformulated in terms of positive numbers.

- Complying with Article 7 GFL zero-tolerances should be followed by further risk assessment.
- The food labelling proposal should not continue in its current form but either be withdrawn or substantially upgraded.

11.2 For administrative authorities

- In exercising powers regard should be had of the legal conditions. The opinion that these conditions are met should be expressed in a reasoned decision.
- CAFAB should comply with Article 9 GFL and Article 41(2) of the EU Charter of Fundamental Rights by engaging stakeholders in deciding on the novelty of food products.
- The European Commission should improve compliance with Article 253 of the EC Treaty and Article 41(2) of the EU Charter of Fundamental Rights by stating in all relevant detail the reasons for exercising its powers.
- Deadlines should be respected.
- Authorities should invest in compliance assistance.

Annexes

Annex 1 Private regulation in the Dutch dairy chain

The first food competitiveness study indicated that self-regulation may add to reducing the problem of administrative burdens resulting from food law. Surprisingly however it also indicated that self-regulation gives rise to substantial investments (Wijnands *et al.*, 2007). The Dairy Private Regulation case study by Maria Litjens set out to shed additional light on the role and impact of private schemes in the Dutch dairy sector.

As mentioned in Wijnands *et al.* (2007), in 2000 the Netherlands' Competition Authority (NMa) set a limit to the use of private standards in the Dutch dairy sector. Some years previously associations of dairy processors and farmers had initiated an integrated quality system for milk, named KKM (Keten Kwaliteit Melk). Processors with a total market share of 98% in the Netherlands and so virtually all dairy farmers as their suppliers, joined this system. In 2000 the NMa refused exemption under the then provision in the Competition Act.[187] The exclusion of non-KKM milk in the agreement of the dairy processors was considered incompatible with the competition law ban on cartels. KKM-requirements went beyond statutory requirements. Dairy farmers producing according to legal requirements should not be denied all market access.[188]

In 2002 the involved organisations made an attempt to have KKM legalised. They convinced the Dutch Dairy Board (Productschap Zuivel) to turn KKM from a private standards into a law-based regulation. Products Boards are compulsory associations of businesses endowed with public regulatory powers.[189] For reasons not relevant in the current context, this regulation was struck down by the courts.[190]

After these setbacks to a collective quality system, dairy processors implemented individual private quality systems. The NMa condoned these systems despite their similarity and their collective excluding effect on non-qualified milk.[191] This cleared the way for a further rise of private food quality arrangements in the Dutch dairy sector. The research underlying the Dairy Private Regulation case study prefers to label these arrangements 'private regulation' rather than 'self-regulation'. The

---

[187] The Dutch Competition Act follows closely the system of European competition law. At EU level Regulation 17/62 provided the option to apply for an exemption under Article 81(3) EC Treaty. Currently Article 81(3) EC Treaty is considered self executing (Regulation 1/2003). This change in approach has been implemented in the Dutch Competition Act as well.
[188] NMa 14 March 2000, case 137.
[189] Product boards are governed by representatives of employers and employees in the sector. They are elements of a corporatist structure of decentralised governance.
[190] CBB 19 May 2004, AB 2004/269.
[191] NMa 14 January 2005, case 4258.

latter label may invoke the image that the regulated party participates voluntarily in the system and partakes in the formulation of the rules. While this picture may conform to contract law theory, it does not to economic reality.

Private regulation relevant for the Dutch dairy sector stretches upstream to the production of ingredients of feed and downstream to distribution to the final consumer. The different activities; single feed production, mixed feed production, farming, processing, distribution, retail and consuming can be seen as inseparable links in the chain. The core elements of the dairy chain in the Netherlands covering products, (association of) businesses and private food quality arrangements are set out below and are presented in Diagram 9.

Raw feed materials
Among the most vulnerable products used on dairy farms is animal feed. The products commonly fed to cattle are known as compound (or mixed) feed. Raw materials for these products are the so-called single feed ingredients, by-products and additives. Single feed ingredients are grain, maize, tapioca, etc. By-products come from the food industry. Single additives as well as premixes (a combination of additives) are also components used in the production of compound feed. In the Netherlands about half of the mixed feed used originates from outside the EU, a quarter from other EU Member States and a quarter is from Dutch origin but may include imported ingredients.

The feed ingredients sector in Europe consists of a considerable number of producers and traders. Many of these are organised in associations. Each association has a *guide to good practice*. More than a dozen of these associations together formed the European Feed Ingredients Platform (EFIP).[192] EFIP is a voluntary platform. It evaluates sectors' guides on the implementation of legal requirements. A benchmark standard has been drawn up to describe the minimum requirements for these guides. One of the member organisations of EFIP, the foundation FAMI-QS owns a sector guide for feed additives and pre-mixtures and uses it in a certification system. The most influential feed quality systems like GMP+ have agreed mutual recognition with FAMI-QS. Certified products in one system are accepted in the other. The FAMI-QS-guide is approved by the European Commission and recognised as a Community Guide to Good Practice.[193]

Compound feed
The next link is the branch of compound feed. Usually a compound feed comprises about twenty different ingredients and a premix. Almost all Dutch compound feed producers are certified by the quality system GMP+ developed by the Product

---

[192] www.efip-ingredients.org.
[193] www.fami-qs.org/index.htm.

Board Animal Feed (Productschap Diervoeder; PDV).[194] For some years GMP+ was implemented in a law-based regulation, but the Product Board changed it to a private law based scheme. Despite its voluntary character, GMP+ to some extent retains its public law image. The branch association of feed manufacturers 'Nevedi' requires in its code of conduct the use of GMP+.[195] GMP+ has been brought in line with other European standards like OVOCOM (Belgium), QS (Germany) and AIC (Great Britain). Together with the European Feed Manufacturers' Federation (FEFAC)[196] the owners of these systems established a joint association, the International Feed Safety Alliance (IFSA), with a common IFIS-standard under construction.[197]

GMP+ incorporates different standards to regulate all kinds of activities in the feed chain: production of feed materials, transport, storage, compound feed processing, etc. However, GMP+ was not considered to be sufficient to assure supply of safe raw materials. For this reason first TrusQ and later Safe Feed as new foundations were established. TrusQ is a closed shop (while foreign manufacturers are welcome to join, it is not open to other Dutch manufacturers beside the initiators). This prompted the establishment of Safe Feed.[198] These two foundations organise control of supply and the suppliers. Individual businesses contribute with capital and labour. The foundations monitor on the basis of GMP+, collect data, etc. for the benefit of the participants. Participants in these schemes may not purchase raw materials from non-complying suppliers.[199]

TrusQ is a co-operative alliance governed by the six feed producers with a total 60 to 80% market share.[200] The aim of TrusQ is to ensure safe feed (and food). The other side of the coin is insurance against safety incidents. A common insurance for the TrusQ-members covers damages related to legal liability with a maximum of € 75 million.[201] Safe Feed by contrast to TrusQ does not organise insurance for its participants, but requires participants to take care of insurance individually.[202] Such insurance should cover damages up to € 2 to 5 million.

Some 200 feed manufacturers supply about 20,000 dairy farmers in the Netherlands. Compound feed represents 25% of total feed, but it is one of the most vulnerable inputs in farm production.

---

[194] www.pdv.nl/index_eng.php?switch=1.
[195] www.nevedi.com/uploads/Gedragscode%20Nevedi.pdf.
[196] www.fefac.org.
[197] www.ifsa-info.net/Engels/index.php?cc=.
[198] www.demolenaar.nl/nieuws/show-nieuws.asp?id=596.
[199] www.trusq.nl/index.php and www.pdv.nl/lmbinaries/08_leverancierbeoordeling_(roordink).pdf.
[200] www.trusq.nl/index.php.
[201] www.trusq.nl/nieuws_details.php?pID=26.
[202] www.safefeed.nl/desktopdefault.aspx.

Milk

Dairy farmers are organised in the Dutch Organisation for Agriculture and Horticulture (Land- en Tuinbouw Organisatie: LTO)[203] and/or the Dutch Dairy Farmers Union (Nederlandse Melkveehouders Vakbond: NMV).[204] LTO is engaged in formulation of general terms and conditions usable for farmers when buying feed. GMP+ certification and an insurance covering to an amount of € 75 million are desired.[205]

In 2006 NMV initiated the Dutch Dairymen Board (DDB), an association intended to negotiate jointly on behalf of the member dairy farmers with processors, because they consider current positions of parties to be such that the milk price is not determined by the market but the processors.[206] DDB signed up to the European Milk Board (EMB) to reach a fair milk price of 40 Cent/kg milk.[207]

Processed milk products

In the Netherlands 12 dairy businesses process virtually[208] all milk. They produce cheese (60%), butter (30%) and consumption milk (10%). Some 60% of the Dutch dairy production is for export, mainly to other EU-countries: Germany, France and Belgium.

After the attempt of the joint quality system KKM failed (as set out above) each processor went for its own private quality system legally constructed as an element of the general terms and conditions or as a recognition scheme linked to an article in the cooperative society constitution.[209] Except Qarant, quality system of Friesland Foods, all are constructed by the Organization Certification Dairy farmers (OCM), successor of the foundation KKM.

In 2005 two processors[210] in the dairy chain (Friesland Foods and Campina, together >80% market share) and a processor in the meat chain (Vion, earlier named Sovion, a company with >80% of the Dutch slaughterhouse capacity) published an intention to require farmers to use only TrusQ or TrusQ-worthy feed.[211] Friesland Foods chose a more general description, but in practice, as can be seen in the model-arrangement for Friesland Foods farmers, only TrusQ-manufacturers can deliver feed to farmers supplying Friesland Foods. In 2007 Campina expressed a

---

[203] www.lto.nl.
[204] www.nmv.nu/sx/index.php.
[205] http://www.gezond-ondernemen.nl/www/modelovereenkomst_veehouder-diervoederleverancier. shtml.
[206] www.ddb.nu.
[207] www.europeanmilkboard.org/en/index.php.
[208] An insignificant quantity is processed on the farm.
[209] www.glipnl and www.frieslandfoods.com.
[210] Now merged into one enterprise.
[211] www.trusq.nl/nieuws_details.php?pID=22, www.frieslandfoods.com/content/zoeken/news. asp?id=2520.

joint approach with six other dairy processors in a letter addressed to Safe Feed. The processors require additional quality guarantees, electronic monitoring of feed and checks of the private insurance of feed manufacturers.[212]

Sale of dairy products
In the Netherlands 98.6 % of dairy products are sold in supermarkets.[213] The Netherlands is a small country, but with a high retail concentration. Three companies account for more than 70% of food sales.[214] All Dutch retailers require their suppliers of retailer-branded products to comply with the BRC[215] food safety standard.[216] The Dutch association of retailers (Centraal Bureau Levensmiddelen: CBL) has required GlobalGAP Integrated Farm Assurance for dairy products as of the first of January 2008.[217] Friesland Foods implements GlobalGAP-requirements in Qarant.

The Global Food Safety Initiative (GFSI), an international cooperation of retailers, has a guide on benchmarking food safety systems.[218] The BRC standard is one of the four recognised certification systems.

Private food quality arrangements in the dairy chain
The arrangements assessed in the Dairy Private Regulation case study set stricter requirements than public law.[219] The links between the arrangements strengthen the integration.

From the Dairy Private Regulation case study follows a categorisation of 3 main types of private schemes (these are represented in the three columns in Diagram 9). The first one expresses food quality arrangements established by one single business. The business establishes corporate standards to be applied by its suppliers. The relation between regulator and regulated is in a vertical direction (continuing lines in Diagram 9). The quality systems are part of general conditions and terms and thus of a contractual relation (Friesland Foods) or linked to the Articles of Association of a cooperative society (Campina) and in this way form part of membership obligations, both with a binding character.

After the legal conflicts around KKM (mentioned above) Friesland Foods chose to create its own private system in order to compete with other businesses. All other businesses followed and have a private scheme, all created by the same independent

---

[212] Letter dated 30-03-2007 of Campina to Safe Feed (referentienr. 2007116/AS/avl.)(unpublished).
[213] www.prodzuivel.nl/pz/productschap/publicaties/sjo/sjo05_engels/SJO_2005_H4.pdf, p. 42.
[214] Connor, 2003, p.3-4; Dobson, 2003.
[215] British Retail Consortium.
[216] Havinga, 2006, p. 528.
[217] Centraal Bureau Levensmiddelen, 2006, p. 13.
[218] www.ciesnet.com/2-wwedo/2.2-programmes/2.2.foodsafety.gfsi.asp.
[219] Information of expert stakeholders and investigation of quality systems.

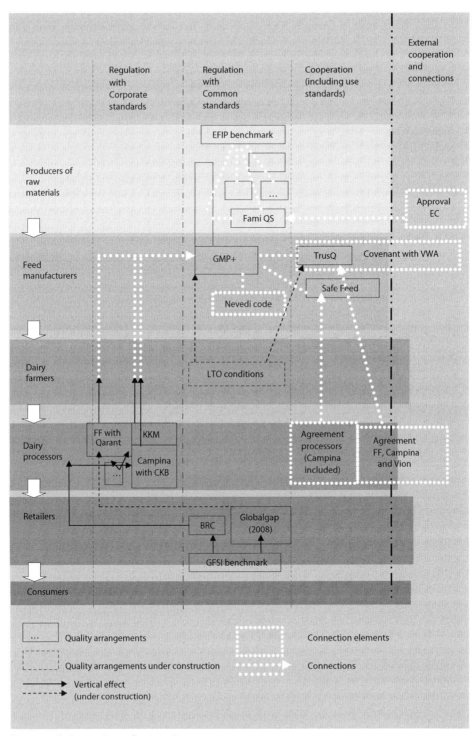

Diagram 9. Dairy chain; food quality arrangement.

foundation. The schemes look similar.[220] For example all include the obligation for farmers to use GMP+ feed. Qarant has this condition too (interrupted lines in figure). Different names seem to be sufficient to satisfy the competition authority.

The second column expresses regulation with collective standards established by company networks, Product Boards or associations of companies. The use of the standards is voluntary. Almost all of these regulations apply in the horizontal as well as the vertical direction by also imposing standards on suppliers (GMP+, LTO, Fami-QS, BRC, and GFSI). Although the private regulation is voluntary, even a non-allied association can apply the regulation as a condition for membership (Nevedi). GMP+ has a number of standards for different sectors and they are geared to each other.

The third column expresses cooperation. Cooperation has different aspects. TrusQ and Safe Feed are foundations with input from and output for individual companies. Public-private cooperation also comes under this column (covenant TrusQ-VWA).

The process of linking food quality arrangements duplicates the effect of the systems (grey interrupted lines in Diagram 9). The connections have different styles.

A quality system can impose an obligation on the regulated business operators to use certified materials (all private systems of dairy processors to farmers hold the obligation to use GMP+ feed). A condition for participation in an association has a comparable effect (Nevedi to GMP+ certification).

Mutual recognition of independent schemes is another style of connection (Fami-Qs and GMP+, GMP+ with Ovocom, QS and AIC). System linking increases the effect of a specific standard at the European or even worldwide level.

A benchmark includes an independent guide.[221] The GFSI benchmark of retailers holds requirements for schemes like BRC. The benchmark standard 'Food Ingredient Standard for sector guides' of EFIP includes only legal requirements, which the sector guides of the member-organisations have to contain.[222]

Connection of arrangements by agreement (Friesland Foods, Campina and Vion) to place obligations on their suppliers is easy to achieve and effective. Connection can have a private-public character by approval. Not only the food quality arrangements are important but also all of the kinds of connections between them.

---

[220] Until 2007 all schemes were available on line.
[221] The term 'benchmark' used in the dairy chain differs from the regular definition: Luning *et al.*, 2002, p. 277.
[222] efip-ingredients.org.dotnet35.hostbasket.com/Feed Ingredient Standard VERSION of 1_12_05.doc.

Annex 2 Novel foods dossier analysis

In the NFR case study by Isabel Cachapa Rodrigues three novel foods dossiers have been analysed. The dossiers have been chosen from the UK because in this country they are made available on the internet. Remarkably in all three a qualification other than novel food would have been defendable. These cases are presented *in extenso* as they shed an interesting light on practices applied in the world of novel foods.

Isomaltulose
On 4 April 2005 the European Commission published a Decision addressed to an applicant (A) authorising the placing on the market of isomaltulose as a novel food or novel food ingredient under Regulation 258/97.[223] On 25 July 2005 another Commission Decision was published addressing applicant (B), also authorising the placing on the market of isomaltulose as a novel food or novel food ingredient under Regulation 258/97.[224]

Isomaltulose[225] is reported to occur naturally at low levels in honey as well as in sugar cane extract.[226] Isomaltulose is produced via enzymatic rearrangement of food-grade[227] sucrose (sugar) using the non-pathogenic bacteria *Protaminobacter rubrum*.[228] In Japan, isomaltulose has been used in yoghurt, chewing gum, and candy since 1985. Annual sales of isomaltulose approximate to 3,000 tons.

Isomaltulose occurs as an intermediate in the production of isomalt (E953), an additive, more precisely a sweetener, permitted for use in food and its production involves the same *Protaminobacter rubrum*. Isomalt is marketed by both applicants A and B. The use of isomalt in food was considered acceptable by the Scientific Committee for Food in 1984.[229] Later an alternative method for making isomalt based on the purification of isomaltulose has been developed and accepted by the Scientific Committee for Food in 1997.[230]

---

[223] Decision 2005/457/EC.

[224] Decision 2005/581/EC.

[225] Isomaltulose is an isomer of sucrose, a reducing disaccharide ($C_{12}H_{22}O_{11}$) produced by an enzymatic conversion of sucrose ($C_{12}H_{22}O_{11}$), whereby the 1.2-glycosidic linkage between glucose and fructose is rearranged to a 1.6-glycosidic linkage. Isomaltulose is a white crystalline substance, characterised by a sweetness quality similar to that of sucrose and a melting point of 123 to 124 °C (Irwin and Sträter, 1991 cited in: Application for the approval of isomaltulose, page 5).

[226] Takazoe, 1985 cited in: Application for the approval of isomaltulose, page 5.

[227] Suitable for human consumption.

[228] The safety of these bacteria was addressed by conducting toxicity studies in mice and rabbits in which live *P. rubrum* cells were administered by intravenous injection. The results of these studies indicated that the organism is nonpathogenic and has a low order of toxigenicity (Porter *et al.*, 1991 cited in Application for the approval of isomaltulose, page 28).

[229] Scientific Committee for Food, 1984.

[230] Scientific Committee for Food, 1997.

Ingested isomaltulose is metabolised in the intestinal mucosa to equal parts glucose and fructose (components of sugar), which are readily absorbed and utilised in carbohydrate metabolic pathways like normal sugar. Besides four separate clinical trials performed by applicant (A) in 2002 and 2003, the toxicological evaluation of isomaltulose was based on a long history of safety studies performed as long as 1963.[231]

The applicant claims that the product is a novel food using the following argument:
'Article 1(2) of EC 258/97 states that the regulation "...shall apply to the placing on the market within the Community of foods and food ingredients which have not hitherto been used for human consumption to a significant degree within the Community..." and which fall under one of six categories of novel foods and novel food ingredients. The lack of a significant prior history of human consumption in the European Community dictates that isomaltulose will be considered under category (c), pertaining to "foods and food ingredients with a new or intentionally modified primary molecular structure". Isomaltulose is thus considered a novel food/food ingredient.'

From the viewpoint of Biochemistry this statement is remarkable. In biochemistry the concept of primary molecular structure refers to the sequence of amino acids in a protein. As regards all other substances, and also simple carbohydrates (such as sugars like sucrose and isomaltulose), one does not refer to them as having a primary molecular structure. Even if one could extrapolate the concept of primary structure from proteins to carbohydrates, then isomaltulose still would not have a modified primary structure in relation to sucrose, because the sequence of glucose and fructose is the same, the only difference is where they are attached.

The applicant further states:
'Section 4 of the Commission Recommendation of 1997 outlines recommendations made by the Scientific Committee for Food (SCF) pertaining to the "Scientific Classification of Novel Foods for the Assessment of Wholesomeness", which facilitates the safety and nutritional evaluation of a given novel food/food ingredient. Of the six classes identified,

---

[231] Sources to this effect quoted by the applicant: (1) Metabolic data in animals (Dahlquist *et al.*, 1963; Goda and Hosoya, 1983; MacDonald and Daniel, 1983; Kawai *et al.*, 1986; Okuda *et al.*, 1986; Tsuji *et al.*, 1986; Ziesenitz, 1986; Goda *et al.*, 1991; Würsch, 1991; Hall and Batt, 1996) and humans (Menzies, 1974; MacDonald and Daniel, 1983; Kawai *et al.*, 1985, 1989; NutriScience, 2003); (2) Clinical data pertaining to the glycaemic response obtained following isomaltulose administration as compared to that obtained with either sucrose or glucose (MacDonald and Daniel, 1983; Kawai *et al.*, 1985, 1989; Liao *et al.*, 2001; NutriScience, 2002); (3) Results of human studies demonstrating that isomaltulose is well-tolerated (MacDonald and Daniel, 1983; Kawai *et al.*, 1985, 1989; Spengler and Sommerauer, 1989); (4) the results of a developmental toxicity study in rats (Lina *et al.*, 1997); and (5) supportive animal sub-chronic and chronic toxicity data, including a study involving feeding of isomaltulose to rats at doses of up to 4,500 mg/kg body weight/day for 26 weeks (Yamaguchi *et al.*, 1986), a 13-week feeding study in rats in which doses up to 8,100 mg/kg body weight were administered, and two additional oral studies focusing particularly on the effect of isomaltulose on tissue mineral content, also conducted in rats, in which doses up to 15,000 mg/kg body weight/day were administered (Kashimura *et al.*, 1990, 1992; Jonker *et al.*, 2002).

isomaltulose would be allocated a Class 1 designation (pure chemicals or simple mixtures from non-GM sources), since it is manufactured by conventional methods as a pure chemical, with no use of genetic modification. Isomaltulose is further classified under sub-class 1.2 (the source of the Novel Food has no history of food use in the Community) of the SCF categorisation. However, isomaltulose occurs as an intermediate product in the production of isomalt (E953), an additive permitted for use in food, which is manufactured by both [...] and [...] and involves use of the same microorganism as that used in the preparation of isomaltulose.'

The applicant considers that '*the source of the Novel Food has no history of food use in the Community*'. This is difficult to accept, after all is not the source of isomaltulose food-grade sucrose? Does not sugar have a history of food use in the Community and for quite some centuries?

If the source is sugar and isomalt is a sweetener it strikes as odd that an intermediate of it would not be a sweetener, as all these substances contain the same glucose and fructose. By consequence, cannot a strong argument be made to qualify isomaltulose as a sweetener, thus as an additive and not as a novel food (additives are outside the scope of the NFR[232])?

Article 1(2) of the Directive on food additives[233] defines food additive as:
'any substance not normally consumed as a food in itself and not normally used as a characteristic ingredient of food whether or not it has nutritive value, the intentional addition of which to food for a technological purpose in the manufacture, processing, preparation, treatment, packaging, transport or storage of such food results, or may be reasonably expected to result, in it or its by-products becoming directly or indirectly a component of such foods.'

Also article 1(2) of the Directive on sweeteners for use in foodstuffs,[234] applies to food additives referred to as sweeteners which are used to '*impart a sweet taste to foodstuffs*' whereas, '*the use of sweeteners to replace sugar is justified for the production of energy-reduced food, noncariogenic foodstuffs or food without added sugars, for the extension of shelf life through the replacement of sugar, and for the production of dietetic products*.' Isomaltulose complies with the definition of food additive and when the applicant in the dossier states the intention to use isomaltulose in beverages and a variety of other products where it would partly replace other sugars as a source of energy, it fulfils also the definition of a sweetener.

Further it is intriguing to note that in its dossier the applicant, when analysing the toxicological aspects of isomaltulose, refers to two documents where isomaltulose

---

[232] Article 2(1)(a) regulation 258/97.
[233] Directive 89/107.
[234] Directive 94/35.

is mentioned. One is from the WHO about malabsorption of food additives.[235] The other is a paper about alternative sweeteners.[236]

This leads to the intermediate conclusion that isomaltulose would better qualify as a sweetener than as a novel food. Why did the applicant apply for authorisation of isomaltulose as a novel food ingredient and not as a food additive? The following facts hint at strategic reasons. Applicant (A) made the novel foods application on 30 October 2003. In the application for approval the applicant mentions the name of a big sugar making company (applicant B), that also markets isomalt (E953) mentioned above and that apparently was the first to prepare isomaltulose as an intermediate compound back in 1957. A comparison of the chemical characteristics and methodology of obtaining the product between both companies is given in the dossier.

Applicant (A) has 7 patents related to isomaltulose, the first was filed in 2001.[237] The other 6 patents are filed in 2005 and 2006 right after the Commission Decision authorising the placing of isomaltulose on the market. Applicant (B) has 39 patents that go from 1986 to 2007.[238] These all refer to a German Patent No. 1049800, published in 1959 that firstly describes a microbiological process for producing crystalline isomaltulose from sucrose using *Protaminobacter rubrum*.[239]

On 4 March 2004 Applicant (B) also made a request for isomaltulose as novel food in Germany. Both applications were evaluated: on 14 February 2005 objections/comments were discussed with Member States for both applications[240] and on 4 April 2005 a decision was reached authorising the placing on the market for Applicant (A) and on 25 July 2005 for Applicant (B).

Applicant (B) made a request for the placing of isomaltulose on the Australian market as a novel food on 27 April 2006. The Australian authority considered isomaltulose to fall within the scope of the definition of 'sugars' as defined in Standard 2.8.1 - Sugars[241] but considered that isomaltulose meets the definition of a 'non-traditional food' in Standard 1.5.1 - Novel Foods as it does not have a history of significant human consumption by the broad community in Australia or New Zealand. The Australian authority also considered that isomaltulose meets the

---

[235] Toxicological versus physiological responses. In: WHO, 1987, p. 82.
[236] Irwin and Sträter, 1991.
[237] http://v3.espacenet.com/results?AB = isomaltulose&sf = a&DB = EPODOC&PA = CERESTAR + &PGS = 10&CY = ep&LG = en&ST = advanced.
[238] http://v3.espacenet.com/results?AB = isomaltulose&sf = a&DB = EPODOC&PA = SUEDZUCKER + & PGS = 10&CY = ep&LG = en&ST = advanced.
[239] http://v3.espacenet.com/textdes?DB = EPODOC&IDX = KR970009295B&F = 0&QPN = KR97000092 95B.
[240] SCFCAH, 2005.
[241] Standard 2.8.1- Sugars. http://www.foodstandards.gov.au/_srcfiles/standard281_sugar_%20v62.pdf.

definition of a 'novel food' based on its composition and structure, and potential patterns and levels of consumption.

Applicant (A) advertises isomaltulose on its website on sweetness solutions where its regulatory status as an accepted novel food is given. This product has a trade mark by Applicant (B).

Trehalose
On 25 September 2001 the Commission published a Decision authorising the placing on the market of trehalose as a novel food or novel food ingredient under Regulation 258/97[242] addressed to applicant (A) on behalf of a Japanese company.

Trehalose is also a naturally occurring simple disaccharide of glucose (sugar) that can be found in bacteria, yeast cells, fungi, algae, a few higher plants, mushrooms, bread, honey and fermented drinks which are consumed as part of an everyday diet. Sugar has properties that are needed when making some foods and trehalose can be used to replace it where a less sweet taste is desired.[243] On the basis of the same arguments that apply to isomaltulose, it is arguable that trehalose falls within the scope of additives – sweeteners.

The Commission in its Decision, mentions that while trehalose, extracted from yeast, was approved for use in foods in 1991[244] in the United Kingdom, it still had to be considered as novel because significant amounts of trehalose had not been marketed in the Member States.[245] No indication had been given by the applicant of whether the trehalose previously approved by the Committee was ever marketed in the Community.

The novel foods application was filed to the competent authority of the United Kingdom on 25 May 2000. The Advisory Committee on Novel Foods and Processes of the UK, had considered in 1990 that when trehalose was added to food it had a technological function and was therefore a food additive, although outside of the then current legislative controls on food additives. The committee was satisfied that there were no food safety concerns.

---

[242] Decision 2001/721.

[243] Application dossier for trehalose.

[244] The Advisory Committee on Novel Foods and Processes (ACNFP) assessed the safety of trehalose, extracted from yeast, in 1990. The intended food applications that were considered by the Committee at that time, included the use of trehalose for the stabilisation of certain foodstuffs during the drying process and upon rehydration (e.g., milk powder, dry soups). On advice of the ACNFP, the Committee on Toxicity (COT), and the Food Advisory Committee (FAC), trehalose was accepted for use in foods (except infant formulae and follow-on formulae) in April 1991. A market for trehalose still had to be established.

[245] In the assessment report of United Kingdom's competent food assessment body it is mentioned that indication was not given by the applicant of whether the trehalose previously approved by the Committee was ever marketed in the Community.

When assessing this application, they stated that trehalose complied with the criteria set by the Joint FAO/WHO Expert Committee on Food Additives JECFA at its 55th meeting (June 6-15, 2000), being considered as safe for use in food.[246] Its functional class was of texturiser, stabiliser, humectant, sweetener. The Advisory Committee on Novel Foods and Processes considered trehalose a food additive, but nevertheless accepted an application for trehalose as a novel food.

According to the applicant trehalose falls under Article 1(2)(f) meaning that it is produced by a novel process, in this case a novel enzymatic process. Four enzymes are used in that process which had not been used hitherto in the EU in food or a food manufacturing process.

How is this claim that trehalose fulfils Article 1(2)(f) to be valued? According to the definition of novel foods and the Commission Recommendation,[247] the product resulting from novel processing is considered to be a novel food only if three conditions are met:
1. where a production process not currently used is applied;
2. giving rise to significant changes in the composition or structure;
3. affecting their nutritional value, metabolism or level of undesirable substances.

Thus although the process used may be novel, this does not automatically make the product a novel food. Let us accept that the process was not used before 15 May 1997 and indeed is new (*condition 1*). The new process has to result in significant changes[248] (*condition 2*); if not then the product is not a novel food. If the food fulfils conditions 1 and 2 it further has to fulfil condition 3 meaning that if there are changes these changes have to affect the criteria of condition 3.

The assessing authority in its opinion[249] states that '*trehalose produced by this enzymatic process is chemically identical to trehalose extracted from yeast*'. If this is true, the production process did not give rise to significant changes that affected its nutritional value metabolism or level of undesirable substances. In that case, trehalose does not meet the definition of a novel food, as it does not it satisfy condition 2.

---

[246] Opinion on an application under the Novel Food Regulation from Bioresco Ltd for clearance of Trehalose produced by a novel enzymatic process. http://www.food.gov.uk/multimedia/pdfs/trehafin. pdf.

[247] Commission Recommendation of 29 July 1997 concerning the scientific aspects and the presentation of information necessary to support applications for the placing on the market of novel foods and novel food ingredients and the preparation of initial assessment reports under Regulation 258/97 of the European Parliament and of the Council (OJ No L253,16.9.97, p. 10).

[248] These changes can be compared either to untreated counterparts or to counterparts which have been processed in a related traditional manner according to the Commission Recommendation on effect of the production process applied to the NF.

[249] http://www.acnfp.gov.uk/assess/fullapplics/trehalose.

Further, '*the Committee agreed that trehalose itself is not a novel product and would have been consumed as a component of a variety of other foodstuffs*'. This confirms that the product is not a novel food.[250] The only aspect assessed as novel was the process.

In fact what the authorities seem to have been mainly evaluating was the quality of the process as they often refer to critical control points (HACCP) that have to be controlled. Evaluating HACCP, however, is not within the scope of the NFR neither does it fall under the safety evaluation of this regulation. Hygiene parameters are not included in the assessment of novel foods[251] but are covered by Regulation 852/2004 of the European Parliament and of the Council of 29 April 2004 on the hygiene of foodstuffs which covers HACCP.

As in the previous case, also in this case one finds a patent. On 4 January 1989 a patent at European level as well as in the USA was filed. This patent was assigned to Quadrant Bioresources, Limited (Cambridge, GB) and concerned 'A method of drying a water-containing foodstuff or beverage at a temperature above ambient, [...] characterized by incorporating trehalose into the foodstuff or beverage which is to be dried.'[252] In Europe the designated contracting states were Belgium, Germany, Spain, France, UK, Italy and the Netherlands among other European countries that at the time of 1997 were not members of the EU.

Ice Structuring Proteins
Ice structuring proteins (ISP) are naturally occurring proteins and peptides, which are found in a variety of living organisms (such as fish, plants, and insects). ISP protects them from damage to tissues in very cold conditions by lowering the temperature at which ice crystals grow and by modifying the size and shape of ice crystals. The applicant intends to market a preparation of ice structuring proteins which will be obtained from the fermentation of genetically modified food grade yeast (*Saccharomyces cervisiae*).[253] This preparation is going to be used in edible ices in order to influence the formation of ice structure during manufacture. This

---

[250] And certainly not in need of a full novel food application procedure. A notification could be justified under article 5 of the NFR based on substantially equivalence with trehalose extracted from yeast.

[251] Commission Recommendation of 29 July 1997 concerning the scientific aspects and the presentation of information necessary to support applications for the placing on the market of novel foods and novel food ingredients and the preparation of initial assessment reports under regulation 258/97 of the European Parliament and of the Council (OJ No L253,16.9.97, p. 11).

[252] European Patent Application publication number 0 297 887 http://v3.espacenet.com/origdoc?DB = EPODOC&IDX = EP0297887&CY = ep&LG = en&QPN = EP0297887.

[253] Decision of the European Commission is still pending: http://ec.europa.eu/food/food/biotechnology/novelfood/app_list_en.pdf.

effect allows, for example, the production of ice cream with a low fat content. The level of ISP in edible ices will not exceed 0.01 % by weight.[254]

At a meeting of the Standing Committee, the issue of the clarification of the status of Ice Structuring Proteins was discussed. A Member State had questioned whether this substance should be regulated as a food additive. There was however more support for the substance to be treated as a novel food. Member States agreed that whatever the framework the important aspect was that the substance would be evaluated for safety.[255]

In line of this event the Advisory Committee on Novel Foods and Processes in the UK in its initial opinion[256] considered ISP to be a novel food ingredient as it had no significant history of consumption in the EU prior to 15 May 1997 and therefore it was considered to fall under category 1(2)(d) that refers to foods and food ingredients consisting of or isolated from microorganisms, fungi or algae.

In the Commission Recommendation such food will be assessed under class 1, pure chemicals or simple mixtures from non-GM sources or class 2, complex NF from non-GM source. ISP, however, is obtained from a GMO and therefore does not fit into these classes.

The safety assessment was performed under class 5.1 for GM microorganisms, the host microorganism used for the genetic modification having a history of use as food or as a source of food in the Community under comparable conditions of preparation and intake. Class 5.1 is associated with novel foods categories related to GMO[257] already removed in 2004,[258] and not with category 1(2)(d).

The safety assessment of ISP, obtained from a GMO, was performed taking into account a guidance document on the risk assessment of products derived from GMO published by EFSA.[259] The scope of this document includes foods produced

---

[254] Opinion on an application under the NFR for ice structuring protein preparation derived from fermented genetically modified baker's yeast *saccharomyces cerevisiae* as a food ingredient. http://www.food.gov.uk/multimedia/pdfs/ispinitialopinion.

[255] SCFCAH, 2006.

[256] Opinion on an application under the NFR for ice structuring protein preparation derived from fermented genetically modified baker's yeast *Saccharomyces cerevisiae* as a food ingredient. http://www.food.gov.uk/multimedia/pdfs/ispinitialopinion.

[257] Categories 1(2)(a) and 1(2)(b).

[258] (a) foods and food ingredients containing or consisting of genetically modified organisms within the meaning of Council Directive 90/220/EEC(2); (b) foods and food ingredients produced from, but not containing, genetically modified organisms. Following this reasoning ISP would fall under 1(2)(b) and not 1(2)(d).

[259] Guidance document of the scientific panel on genetically modified organisms for the risk assessment of genetically modified plants and derived food and feed, May 2006. http://www.efsa.europa.eu/EFSA/Scientific_Document/gmo_guidance_gm_plants_en.pdf.

using genetically modified microorganisms irrespective of whether they do not, or do fall under Regulation 1829/2003.[260]

Why was ISP, obtained by GMO, considered to fall under a category of novel food related to non-GM source while its safety assessment was conducted according to categories related to GM sources? It seems that ISP was artificially fitted into the scope of this regulation. This could reflect what was mentioned previously; that Member States considered that the important aspect was that the substance would be evaluated for safety. Apparently, the Member States wanted the safety of ISP to be assessed whatever the framework would be.

The framework chosen was that of the NFR. Yet another qualification was again defendable. This other qualification is that of ISP being a processing aid.[261] ISP performs its major technological function during manufacture of ice cream and edible ice products (where it alters the growth and shape of developing ice crystals and hence the ice structure and properties) and has no technological function in the final food according to the definition.

This novel foods application was made in 2006. Remarkably, the applicant had requested in 2004 to the Australian authorities, authorisation for placing ISP on the Australian market as a processing aid. Further in Australia, the applicant considered that ISP protein was not a novel protein since it is identical to that obtained from a fish which is consumed as food. In Australia the definition of processing aid[262] is comparable with the European one if not stricter.

In its final opinion[263] the Australian authority approved ISP as a processing aid since '*it performs its technological function during the manufacture of ice cream and edible ices and does not have a food additive function in the final food. Its use for the*

---

[260] Regulation 1829/2003.

[261] For the purpose of Council Directive of 21 December 1988 on the approximation of the laws of the Member States concerning food additives authorized for use in foodstuffs intended for human consumption (89/107/EEC) 'processing aid' means any substance not consumed as a food ingredient by itself, intentionally used in the processing of raw materials, foods or their ingredients, to fulfill a certain technological purpose during treatment or processing and which may result in the unintentional but technically unavoidable presence of residues of the substance or its derivatives in the final product, provided that these residues do not present any health risk and do not have any technological effect on the finished product. http://eur-lex.europa.eu/LexUriServ/LexUriServ.do?uri=CONSLEG:1989L01 07:20031120:EN:PDF.

[262] Processing aid is defined as: (a) the substance is used in the processing of raw materials, foods or ingredients, to fulfil a technological purpose relating to treatment or processing, but does not perform a technological function in the final food; and (b) the substance is used in the course of manufacture of a food at the lowest level necessary to achieve a function in the processing of that food, irrespective of any maximum permitted level specified. (Standard 1.3.3 of the Australia New Zealand Food Standards Code), http://www.foodstandards.gov.au/_srcfiles/Standard_1_3_3_Processing_Aids_v94.doc.

[263] Final assessment report, application A544, Ice structuring protein as a processing aid for ice cream & edible ices, http://www.foodstandards.gov.au/_srcfiles/FAR_A544%20ISP%20FAR%20FINAL.pdf.

proposed purpose is technologically justified as a processing aid.' Also the Australian authority commented that *'ISP does not contain any novel DNA or novel protein since it is nature identical to a naturally occurring fish protein consumed in the diet.'*

Why did the applicant make a request in Australia for ISP as a processing aid mentioning that it is not novel and in the EU applied for the full authorisation as novel food? Alternatively the applicant could also in the EU take the position that ISP is a processing aid or at least go no further than to submit a notification under Article 5 NFR on basis of substantially equivalence with the protein obtained from a fish which is consumed as food, as argued in Australia.

Also in this case there are patents related to ISP that start from 2004 until 2007.[264]

Businesses' identifications of foods as novel tend to be followed by authorities.

The conclusion from these cases is that businesses in applying for premarket approval, choose between the different schemes. Authorities seem to follow the choice made regardless of the conformity of the choice to the legal framework.

All cases show applicants apparently keen on acquiring protection for their products as in all cases they have acquired patents. This may indicate that the apparent preference of these applicants for the Novel Foods framework over other frameworks, is related to the exclusive nature of the authorisation of Novel Foods as opposed to the other authorisation types.

---

[264] http://v3.espacenet.com/results?AB = ice + structuring + proteins&sf = a&DB = EPODOC&PGS = 10 &CY = ep&LG = en&ST = advanced.

Annex 3 Abbreviations

AB	Nederlandse jurisiprudentie Administratiefrechtelijke Beslissingen (Dutch Review of case law in administrative law)
ADI	Acceptable Daily Intake
AIC	Agricultural Industries Confederation
AROO	WTO Agreement on Rules of Origin
BRC	British Retail Consortium
BSE	Bovine Spongiform Encephalopathy
CAFAB	Novel Foods Working Group (Competent Authority Food Assessment Body)
CBL	Dutch Food Retail Association
CCPs	Critical Control Points
CIAA	Confederation of the Food and Drink Industries of the European Union
CIES	Comité International d'Entreprises à Succursales (International Committee of Food Retail Chains)
CO	Country of Origin
COOL	Country of Origin Labelling
CRO	Committee on Rules of Origin
DDB	Dutch Dairymen Board
DG	Directorate General
EC	European Community
ECJ	European Court of Justice
ECR	European Court Reports
EDA	European Dairy Association
EFIP	European Feed Ingredients Platform
EFFL	European Food & Feed Law Review
EFLA	European Food Law Association
EFSA	European Food Safety Authority
EFTA	European Free Trade Association
EMB	European Milk Board
(DG) ENTR	DG Enterprise and Industry
FAO	Food and Agriculture Organisation of The United Nations
FAMI-QS	European Feed Additives and PreMixtures Quality System
FDA	U.S. Food and Drug Administration
FEFAC	European Feed Manufacturers' Federation
FTA	Free Trade Area
FEFAC	European Feed Manufacturers' Federation
GAP	Good Agricultural Practices
GATT	General Agreement on Tariffs and Trade
GFL	General Food Law (Regulation 178/2002)
GFSI	Global Food Safety Initiative
GHI	Global Harmonization Initiative

GM   Genetically Modified
GMOs   Genetically Modified Organisms
GMP   Good Manufacturing Practices
HACCP   Hazard Analysis and Critical Control Point
IFIS   International Feed Ingredients Standard
IFS   International Food Standard
IFSA   International Feed Safety Alliance
ISO   International Organization for Standardization
ISP   Ice Structuring Proteins
JECFA   Joint FAO/WHO Expert Committee on Food Additives
JEMRA   Joint FAO/WHO Meetings on Microbiological Risk Assessment
JMPR   Joint FAO/WHO Meetings on Pesticide Residues
KKM   Integrated Quality System for Milk (Keten Kwaliteit Melk)
LTO   Dutch Organisation for Agriculture and Horticulture (Land-en
            Tuinbouw Organisatie)
MC   Member Countries
MLK   Münster´sches Lebensmittelrechts Kolloquium
MRL   Maximum Residue Level
NFR   Regulation 258/97 concerning Novel Foods and Novel Food Ingredients
NMa   Dutch Competition Authority
NMV   Dutch Dairy Farmers Union (Nederlandse Melkveehouders Vakbond)
NNFR   Proposal for a Regulation of the European Parliament and of the
            Council on novel foods and amending Regulation XXX/XXXX
            [common procedure] (COM(2007) 872)
NOAEL   No Observed Adverse Effects Level
NVLR   Nederlandse Vereniging voor Levensmiddelenrecht
OCM   Organisation Certification Dairy farmers
OJ   Official Journal of the European Union
OM   Origin Marking
OVOCOM   Concertation Platform for Feed
PDO   Protected Designation of Origin
PDV   Dutch Product Board Animal Feed (Productschap Diervoeder)
PGI   Protected Geographical Indication
PIE   Platform for Ingredients in Europe
Ppb   Parts per billion
QS   Qualität und Sicherheit
REACH   Registration, Evaluation, Authorisation and Restriction of Chemical
            substances (Regulation 1907/2006)
ROOs   Rules of Origin
(DG) Sanco DG Health and Consumer Protection
SCF   Scientific Committee for Food
SCFCAH   Standing Committee on the Food Chain and Animal Health
SME   Small and medium-sized enterprise

SPS (Agreement)   (WTO Agreement on the Application of) Sanitary and
            Phytosanitary Measures
SQF   Safe Quality Food
TBT (Agreement)   (WTO Agreement on) Technical Barriers to Trade
TCRO   Technical Committee on Rules of Origin
TOR   Terms of reference
UEAPME   Union européenne de l´artisanat et des petites et moyennes entreprises
v.   versus (against)
VWA   Dutch Food Safety Authority (Voedsel en Waren Autoriteit)
WCO   World Customs Organisation
WHO   World Health Organisation
WTO   World Trade Organisation
ZLR   Zeitschrift für das gesamte Lebensmittelrecht

References

Advisory Committee on Novel Foods and Processes (ACNFP), 1990. Annual report 1990. Available at: http://www.food.gov.uk/multimedia/pdfs/acnfp9001.pdf.

Brookes, G., 2007. Briefing Report: Economic impact assessment of the way in which the EU novel foods regulatory approval procedures affect the EU food sector-for the Confederation of the food and drink industries of the European Union (CIAA) & the Platform for ingredients in Europe (PIE). Available at: http://www.ciaa.be/documents/news_events/finalreport2july2007.pdf.

Bremmers, H., Van der Meulen, B., Poppe, K. and Wijnands, J., 2008. Administrative burdens in the European Food Industry – with special attention to the dairy sector. LEI report 2008-066, Available at: http://www.lei.dlo.nl/publicaties/PDF/2008/2008-066.pdf.

CBB 19 May 2004. AB 2004/269.

COM(2005) 661 Proposal for a Council Regulation on the indication of the country of origin of certain products imported from third countries. Available at: http://eur-lex.europa.eu/LexUriServ/LexUriServ.do?uri = COM:2005:0661:fin:en:pdf.

COM(2007) 23 Action programme for reducing administrative burdens in the European Union. Available at: http://eur-lex.europa.eu/LexUriServ/LexUriServ.do?uri = COM:2007:0023:fin:en:pdf.

COM(2007) 90 Proposal for a Regulation of the European Parliament and of the Council amending Regulation No 11 concerning the abolition of discrimination in transport rates and conditions, in implementation of Article 79(3) of the Treaty establishing the European Economic Community and Regulation (EC) No 852/2004 of the European Parliament and the Council on the hygiene of foodstuffs. Available at: http://eur-lex.europa.eu/LexUriServ/LexUriServ.do?uri = COM:2007:0090:fin:en:pdf.

COM(2007) 469 Report from the Commission to the European Parliament and the Council on existing legal provisions, systems and practices in the Member States and at Community level relating to liability in the food and feed sectors and on feasible systems for financial guarantees in the feed sector at Community level in accordance with Article 8 of Regulation (EC) No 183/2005 of the European Parliament and of the Council of 12 January 2005 laying down requirements for feed hygiene. Available at: http://eur-lex.europa.eu/LexUriServ/LexUriServ.do?uri = COM:2007:0469:fin:en:pdf.

COM(2007) 872 Proposal for a Regulation of the European Parliament and of the Council on novel foods and amending Regulation (EC) No XXX/XXXX [common procedure]. Available at: http://eur-lex.europa.eu/LexUriServ/LexUriServ.do?uri = COM:2007:0872:fin:en:pdf.

COM(2008) 40 Proposal for a Regulation of the European Parliament and of the Council on the provision of food information to consumers. Available at: http://eur-lex.europa.eu/LexUriServ/LexUriServ.do?uri = COM:2008:0040:fin:en:pdf.

COM(2008) 824 Report from the Commission to the Council and the European Parliament on the use of substances other than vitamins and minerals in food supplements. Available at: http://eur-lex.europa.eu/LexUriServ/LexUriServ.do?uri = COM:2008:0824:fin:en:pdf.

References

Connor, J.M., 2003. The changing structure of global food markets: dimensions, effects, and policy implications. In: OECD Conference on changing dimensions of the food economy: exploring the policy issues, 6-7 February 2003, The Hague, the Netherlands. Available at: http://www.oecd.org/document/47/0,3343,en_2649_33781_20175727_1_1_1_1,00.html.

Curtis, P.A., 2005. Guide to food laws and regulations. 1st ed. Blackwell Publishing Profesional, Ames, Iowa, USA.

Dahlquist, A., Auricchio, S.; Semenza, G. and Prader, A., 1963. Human intestinal disaccharidases and hereditary disaccharide intolerance. The hydrolysis of sucrose, isomaltose, palatinose (isomaltulose), and a 1,6-α-oligosaccharide (isomaltoligosaccharide) preparation. Journal of Clinical Investigation 42: 556-562.

Decision 2001/721/EC. Commission Decision of 25 September 2001 authorising the placing on the market of trehalose as a novel food or novel food ingredient under Regulation (EC) No 258/97 of the European Parliament and of the Council (notified under document number C(2001) 2687). Official Journal of the European Union L 269: 17-19. Available at: http://eur-lex.europa.eu/LexUriServ/LexUriServ.do?uri=OJ:L:2001:269:0017:0019:en:pdf.

Decision 2005/457/EC. Commission Decision of 4 April 2005 authorising the placing on the market of isomaltulose as a novel food or novel food ingredient under Regulation (EC) No 258/97 of the European Parliament and of the Council (notified under document number C(2005) 1001). Official Journal of the European Union L 160: 28-30. Available at: http://eur-lex.europa.cu/LexUriServ/LexUriServ.do?uri=OJ:L:2005:160:0028:0030:en:pdf.

Decision 2005/581/EC. Commission Decision of 25 July 2005 authorising the placing on the market of isomaltulose as a novel food or novel food ingredient under Regulation (EC) No 258/97 of the European Parliament and of the Council (notified under document number C(2005) 2776). Official Journal of the European Union L 199: 90-91. Available at: http://eur-lex.europa.eu/LexUriServ/LexUriServ.do?uri=OJ:L:2005:199:0090:0091:en:pdf.

DG Sanco, 2006. Labelling: competitiveness, consumer information and better regulation for the EU (No identification number). Available at: http://ec.europa.eu/food/food/labellingnutrition/betterregulation/competitiveness_consumer_info.pdf.

DG Trade, 2003. Made in the EU origin marking – working document of the Commission Services, Brussels, 12 December 2003. Available at: http://trade.ec.europa.eu/doclib/docs/2005/may/tradoc_115557.pdf.

DG Trade, 2004. Consideration of an EU origin marking scheme, consultation process, analysis and next steps. Available at: http://trade.ec.europa.eu/doclib/docs/2005/may/tradoc_118123.pdf.

DG Trade, 2005. Impact assessment, Commission staff working document, annex to the Proposal for a Council Regulation on the indication of the country of origin of certain products imported from third countries, Brussels, 16.12.2005 SEC(2005) 1657. Available at: http://trade.ec.europa.eu/doclib/docs/2005/december/tradoc_126710.pdf.

DG Trade, 2006a. European Commission (directorate-general for trade), 'Made-in'– a EU origin marking scheme, parameters and prospects, Brussels, 13 January 2006. Available at: http://trade.ec.europa.eu/doclib/docs/2006/january/tradoc_127021.pdf.

DG Trade, 2006b. Origin marking, state of play and orientations for the way forward. Available at: http://trade.ec.europa.eu/doclib/docs/2006/july/tradoc_129338.pdf.

Directive 89/107. Council Directive of 21 December 1988 on the approximation of the laws of the Member States concerning food additives authorized for use in foodstuffs intended for human consumption (89/107/EEC). Amended by European Parliament and Council Directive 94/34/EC of 30 June 1994 amending Directive 89/107/EEC on the approximation of the laws of Member States concerning food additives authorized for use in foodstuffs intended for human consumption. Amended by Regulation (EC) No 1882/2003 of the European Parliament and of the Council of 29 September 2003 adapting to Council Decision 1999/468/EC the provisions relating to committees which assist the Commission in the exercise of its implementing powers laid down in instruments subject to the procedure referred to in Article 251 of the EC Treaty. Available at: http://eur-lex.europa.eu/LexUriServ/site/en/consleg/1989/L/01989L0107-20031120-en.pdf.

Directive 94/35. European Parliament and Council Directive 94/35/EC of 30 June 1994 on sweeteners for use in foodstuffs to impart a sweet taste to foodstuffs. Official Journal of the European Union L 237: 3-12. Available at: http://eur-lex.europa.eu/LexUriServ/LexUriServ.do?uri=CELEX:31994L0035:en:html.

Directive 2000/13. Directive 2000/13/EC of the European Parliament and of the Council on the approximation of the laws of the Member States relating to the labelling, presentation and advertising of foodstuffs. Official Journal of the European Union L 109: 29-42. Available at: http://eur-lex.europa.eu/LexUriServ/LexUriServ.do?uri=OJ:L:2000:109:0029:0042:EN:PDF.

Directive 2008/5. EC of 30 January 2008 concerning the compulsory indication on the labelling of certain foodstuffs of particulars other than those provided for in Directive 2000/13/EC of the European Parliament and of the Council. Official Journal of the European Union L 27: 12-16. Available at: http://eur-lex.europa.eu/LexUriServ/LexUriServ.do?uri=OJ:L:2008:027:0012:0016:en:pdf.

Dobson, P.W., 2003. Buyer power in food retailing: The European experience, In: OECD Conference on changing dimensions of the food economy: exploring the policy issues, 6-7 February 2003, The Hague, the Netherlands. Available at: http://www.oecd.org/document/47/0,3343,en_2649_33781_20175727_1_1_1_1,00.html.

European Court of Justice 11 July 1974. Procureur du Roi v. Dassonville, case 8-74. ECR 1974, p. 00837.

European Court of Justice 25 April 1985. Commission v. UK, case 207/83. ECR 1985, p. 01201.

European Court of Justice 5 November 2002. Commission v. Germany, case C-325/00. ECR 2002, I-09977.

European Court of Justice 6 March 2003. Commission v. Germany, case C-6/02. ECR 2003, I-02389.

European Court of Justice 23 September 2003. Commission v. Denmark, case 192/01. ECR 2003, I-9693.

European Court of Justice 5 February 2004. Commission v. French Republic, case C-24/00. ECR 2004 para. 26.

European Court of Justice 12 July 2005. The Queen, on the application of alliance for natural health and Nutri-Link Ltd v Secretary of State for Health (C-154/04) and The Queen, on the application of National Association of health stores and Health Food Manufacturers Ltd v Secretary of State for Health and National Assembly for Wales (C-155/04), joined cases C-154/04 and C-155/04. ECR 2005, I-06451.

European Court of Justice 28 September 2006. NV Raverco (C-129/05), Coxon & Chatterton Ltd (C-130/05) v. Minister van Landbouw, Natuur en Voedselkwaliteit (in Dutch), joined cases C-129/05 and C-130/05.

EFTA Court Case E-3/00, EFTA Surveillance Authority v. Norway, EFTA Court Report 2000/01.

European Commission, 1997. Commission Recommendation of 29 July 1997 concerning the scientific aspects and the presentation of information necessary to support applications for the placing on the market of novel foods and novel food ingredients and the preparation of initial assessment reports under regulation (EC) No 258/97 of the European Parliament and of the Council, 97/618. Official Journal of the European Union L 253. Available at: http://eur-lex.europa.eu/LexUriServ/LexUriServ.do?uri=celex:31997h0618:en:html.

European Commission, 2000a. White paper on food safety. Available at: http://ec.europa.eu/dgs/health_consumer/library/pub/pub06_en.pdf.

European Commission, 2000b. Commission Communication on the precautionary principle Brussels 2.2.2000 COM(2000) 1 final. Available at: http://ec.europa.eu/dgs/health_consumer/library/pub/pub07_en.pdf.

European Commission, 2003. Questions and answers on residues and contaminants in foodstuffs, Brussels, 19 February 2003. Available at: http://ec.europa.eu/food/food/chemicalsafety/residues/fcr_qanda_en.pdf.

European Commission, 2004. Report from the Commission to the Council and the European Parliament on the implementation of Title II of Regulation (EC) No 1760/2000 of the European Parliament and of the Council establishing a system for the identification and registration of bovine animals and regarding the labelling of beef and beef products, Brussels, 27.04.2004 COM(2004)316 final. Available at: http://www.defra.gov.uk/foodrin/beeflab/ec_beeflabelling.pdf.

European Community, 2005. Comments from the European Community on veterinary drugs without ADI/MRL, Codex Committee on residues of veterinary drugs in foods 22/04/2005, (CL 2004/50-RVDF).

FAO, 2005. Voluntary guidelines, to support the progressive realization of the right to adequate food in the context of national food security, Rome 2005. Available at: http://www.fao.org/docrep/meeting/009/y9825e/y9825e00.htm.

Food Standards Agency, 2000. Application dossier for trehalose. Available at: http://www.food.gov.uk/multimedia/pdfs/0_1.pdf.

Food Standards Agency, 2003. Application for the approval of isomaltulose. Available at: http://www.food.gov.uk/multimedia/pdfs/isomaltulose.pdf.

Goda, T. and Hosoya, N., 1983. Hydrolysis of palatinose by rat intestinal sucrase-isomaltase complex. Nihon Eiyo Shokuryo Gakkaishi 36: 169-173. Cited in: Würsch, P. 1991. Metabolism and tolerance of sugarless sweeteners. In: Rugg-Gunn, A.J. (Ed.). Sugarless: the way forward. Elsevier Applied Science; New York, USA, pp. 32-51.

Goda, T., Takase, S. and Hosoya, N., 1991. Hydrolysis of palatinose condensates by rat intestinal disaccharidases. Nihon Eiyo Shokuryo Gakkaishi 44: 395-398.

Grit, C. and Tiesinga, I., 2008. Europees voorstel Voedselinformatie; betere wet- en regelgeving, of toch niet? (in Dutch) Journaal Warenwet 144: 7-13.

Hagenmeyer, M., 2008. The regulation overkill: food information, European Food and Feed Law review, 3/3/2008, pp. 165-171.

Hall, E.J. and Batt, R.M., 1996. Urinary excretion by dogs of intravenously administered simple sugars. Research in Veterinary Science 60: 280-282.

Havinga, T., 2006. Private regulation of food safety by supermarkets. Law & Policy 28: 515-533. Available at: http://papers.ssrn.com/sol3/papers.cfm?abstract_id=929001.

Hoogland, A. i.s.m. branchevereniging NPN, 2006. Novel Food-wetgeving. Protectionisme of consumentenbescherming? Available at: http://magazine.vannature.nl/downloads/20062/artikel_novel_food_wetgeving.pdf.

Irwin, W.E. and Sträter, P.J., 1991. Isomaltulose. In: O'Brien Nabors, L. and Gelardi, R.C. (eds.). Alternative sweeteners (2nd Rev. Expanded Ed.), Marcel Dekker; New York, pp. 299-307.

Jonker, D., Lina, B.A.R. and Kozianowski, G. 2002. 13-week oral toxicity study with isomaltulose (Palatinose®) in rats. Food and Chemical Toxicology 40: 1383-1389.

Kashimura, J., Kimura, M., Kondo, H., Yokoi, K., Nishio, K., Nakajima, Y. and Itokawa, Y. 1990. Effects of feeding of Palatinose® and its condensates on tissue mineral contents in rats. Nihon Eiyo Shokuryo Gakkaishi 43:127-131.

Kashimura, J., Kimura, M., Kondo, H., Yokoi, K., Nakajima, Y., Nishio, K. and Itokawa, Y. 1992. Effects of Palatinose and its condensates on contents of various minerals in rat various tissues. Nihon Eiyo Shokuryo Gakkaishi 45:49-54.

Kawai, K., Okuda, Y. and Yamashita, K. 1985. Changes in blood glucose and insulin after an oral palatinose administration in normal subjects. Endocrinologia japonica 32: 933-936.

Kawai, K., Okuda, Y., Chiba, Y. and Yamashita, K. 1986. Palatinose as a potential parenteral nutrient: its metabolic effects and fate after oral and intravenous administration to dogs. Journal of Nutritional Science and Vitaminology 32: 297-306.

Kawai, K., Yoshikawa, H., Murayama, Y., Okuda, Y. and Yamashita, K. 1989. Usefulness of palatinose as a caloric sweetener for diabetic patients. Hormone and Metabolic Research 21: 338-340.

Kozlovski, R.O. and Van der Kroon, T.S., 2008. Hercodificatie etikettering: wordt het etiket nog voller? Etiketteringsvoorstellen ontmoeten scepsis, (in Dutch) Journaal Warenwet 4/9/2008, 133, pp. 56-58.

Krissoff, B., Kuchler, F., Nelson, K., Perry, J. and Somwaru, A., 2004. Country-of-Origin Labeling: Theory and Observation. Outlook Report from the Economic research Service, USDA, USA. Available at: http://www.ers.usda.gov/publications/WRS04/jan04/wrs0402.

Lang, S. and Gaisford, J., 2007. Rules of origin and tariff circumvention, chapter 12. In: W.A. Kerr and J.D. Gaisford (eds.), Handbook on international trade policy, Edward Elgar, Cheltenham, UK.

Lelieveld, H. and Keener L., 2007. Global harmonization of food regulations and legislation – the Global Harmonization Initiative. Trends in Food Science & Technology 18: S15-S19; Available at: http://www.icc.or.at/international_research/2007_EHEDG_Yearbook-TIFS18.pdf.

Liao, Z.-H., Li, Y.-B., Yao, B., Fan, H.-D., Hu, G.-L. and Weng, J.-P. 2001. The effects of isomaltulose on blood glucose and lipids for diabetic subjects. Diabetes 50 (Suppl. 2): A366 [Abstract No. 1530-P].

Lina, B.A., Smits-Van Prooije, A.E. and Waalkens-Berendsen, D.H. 1997. Embryotoxicity/teratogenicity study with isomaltulose (Palatinose®) in rats. Food and Chemical Toxicology 35: 309-314.

Luning, P.A., Marcelis, W.J. and Jongen, W.M.F., 2002. Food quality management. A techno-managerial approach. Wageningen Academic Publishers, Wageningen, the Netherlands.

MacDonald, I. and Daniel, J.W. 1983. The bioavailability of isomaltulose in man and rat. Nutrition Reports International 28: 1083-1090.

Menzies, I.S., 1974. Absorption of intact oligosaccharide in health and disease. Biochemical Society Transactions 2: 1042-1047.

Mettke, T., 2008. Die Überspannung der europäischen Lebensmittelgesetzgebung, ZLR 3/35/2008: 381-389.

MRA EU-USA, 1998. Agreement on mutual recognition between the European Community and the United States of America. Available at: http://www.ustr.gov/assets/World_Regions/Europe_Middle_East/Europe/1998_US-EU_Mutual_Recognition_Agreement/asset_upload_file292_7083.pdf.

NMa 14 March 2000, case 137. Available at: www.nmanet.nl.

NMa 14 January 2005, case 4258, Informele zienswijze borgingsysteem kwaliteit productie melk. Available at: www.nmanet.nl.

NutriScience, 2002. The effect of dextrose and isomaltulose ingestion on serum glucose and insulin levels in healthy volunteers. NutriScience Report 72.01.0003.

NutriScience, 2003. Study on the intestinal absorption of isomaltulose, trehalose, and soy-isoflavones. Report on isomaltulose. NutriScience Report 72.01.0010/B.

O'Connor, B., 2004. The law of geographical indications. Cameron May, London, UK (reprinted 2007).

Okuda, Y., Kawai, K., Chiba, Y., Koide, Y. and Yamashita, K. 1986. Effects of parenteral palatinose on glucose metabolism in normal and streptozotocin diabetic rats. Hormone and Metabolic Research 18: 361-364.

Ortega Medina, 1997. Report on alleged contraventions or maladministration in the implementation of Community law in relation to BSE, 7 February 1997 A4-0020/97/A. Available at: http://www.mad-cow.org/final_EU.html.

Poppe, K.J., Wijnands, J.H.M., Bremmers, H.B.M., Van der Meulen, B.M.J. and Tacken, G.L., 2009. Food legislation and competitiveness in the EU food industry. Case studies in the dairy industry (end report) European Communities 2009. Available at: http://ec.europa.eu/enterprise/sectors/food/files/competitiveness/food_legislation_dairy_sector_en.pdf.

Porter, M.C., Kuijpers, M.H.M., Mercer, G.D., Hartnagel, R.E. (Jr.), and Koeter, H.B.W.M., 1991. Safety evaluation of *Protaminobacter rubrum*: intravenous pathogenicity and toxigenicity study in rabbits and mice. Food and Chemical Toxicology 29: 685-688, Cited in: Application for the approval of isomaltulose, page 28, Available at: http://www.food.gov.uk/multimedia/pdfs/isomaltulose.pdf.

Rechtbank – district court – Leeuwarden 28 May 2004. LJN AP1271; LJN is the identification of cases used on www.rechtspraak.nl. For a summary in English see: EFFL 3/2007, p. 171.

Rechtbank – court of appeal – Leeuwarden 30 November 2006. LJN AZ 3591; LJN is the identification of cases used on www.rechtspraak.nl. For a summary in English see: EFFL 3/2007, p. 171.

Regulation 2913/92. Regulation (EC) establishing the Community Customs Code. Official Journal of the European Union L 302: 1-50. Available at: http://eur-lex.europa.eu/LexUriServ/LexUriServ.do?uri=CELEX:31992R2913:en:html.

Regulation 2200/96. Regulation (EC) on the common organization of the market in fruit and vegetables. Official Journal of the European Union L 297: 1-28. Available at: http://eur-lex.europa.eu/LexUriServ/LexUriServ.do?uri=CELEX:31996R2200:en:html.

Regulation 258/97. Regulation (EC) of the European Parliament and of the Council of 27 January 1997 concerning novel foods and novel food ingredients. Official Journal of the European Union L 43: 1. Available at: http://eur-lex.europa.eu/LexUriServ/LexUriServ.do?uri=CONSLEG:1997R0258:20040418:en:pdf.

Regulation 178/2002. Regulation (EC) of the European Parliament and of the Council of 28 January 2002 laying down the general principles and requirements of food law, establishing the European Food Safety Authority and laying down procedures in matters of food safety. Official Journal of the European Union L 31: 1. Available at: http://eur-lex.europa.eu/LexUriServ/LexUriServ.do?uri=OJ:L:2002:031:0001:0024:en:pdf.

Regulation 1829/2003. Regulation (EC) of the European Parliament and of the Council of 22 September 2003 on genetically modified food and feed. Official Journal of the European Union L 268: 1. Available at: http://eur-lex.europa.eu/LexUriServ/LexUriServ.do?uri=OJ:L:2003:268:0001:0023:en:pdf.

Regulation 853/2004. Regulation (EC) of the European Parliament and of the Council laying down specific hygiene rules for food of animal origin. Official Journal of the European Union L 139: 55-205. Available at: http://eur-lex.europa.eu/LexUriServ/LexUriServ.do?uri=OJ:L:2004:139:0055:0205:en:pdf.

Regulation 854/2004. Regulation (EC) of the European Parliament and of the Council laying down specific rules for the organisation of official controls on products of animal origin intended for human consumption. Official Journal of the European Union L 226: 83-127. Available at: http://eur-lex.europa.eu/LexUriServ/LexUriServ.do?uri=OJ:L:2004:226:0083:0127:EN:PDF.

Regulation 882/2004. Regulation (EC) of the European Parliament and of the Council of 29 April 2004 on official controls performed to ensure the verification of compliance with feed and food law, animal health and animal welfare rules. Official Journal of the European Union L 191: 1-52 (corrigendum). Available at: http://eur-lex.europa.eu/LexUriServ/LexUriServ.do?uri=celex:32004r0882r(01):en:html.

Regulation 396/2005. Regulation (EC) on maximum residue levels of pesticides in or on food and feed of plant and animal origin and amending Council Directive 91/414/EEC. Official Journal of the European Union L 70: 1. Available at: http://eur-lex.europa.eu/LexUriServ/LexUriServ.do?uri=OJ:L:2005:070:0001:0016:en:pdf.

Regulation 2074/2005. Regulation (EC) laying down implementing measures for certain products under Regulation (EC) No 853/2004 of the European Parliament and of the Council and for the organisation of official controls under Regulation (EC) No 854/2004 of the European Parliament and of the Council and Regulation (EC) No 882/2004 of the European Parliament and of the Council, derogating from Regulation (EC) No 852/2004 of the European Parliament and of the Council and amending Regulations (EC) No 853/2004 and (EC) No 854/2004. Official Journal of the European Union L 338: 27. Available at: http://eur-lex.europa.eu/LexUriServ/LexUriServ.do?uri=OJ:L:2005:338:0027:0059:en:pdf.

Regulation 1182/2007. Regulation (EC) laying down specific rules as regards the fruit and vegetable sector. Official Journal of the European Union L 273: 1-30. Available at: http://eur-lex.europa.eu/LexUriServ/LexUriServ.do?uri=OJ:L:2007:273:0001:0030:en:pdf.

Regulation 1234/2007. Regulation (EC) establishing a common organisation of agricultural markets and on specific provisions for certain agricultural products (Single CMO Regulation). Official Journal of the European Union L 299: 1-149. Available at: http://eur-lex.europa.eu/LexUriServ/LexUriServ.do?uri=OJ:L:2007:299:0001:0149:EN:PDF.

Regulation 1331/2008. Regulation (EC) establishing a common authorisation for food additives, food enzymes and food flavourings. Official Journal of the European Union L 354: 1-6. Available at: http://eur-lex.europa.eu/LexUriServ/LexUriServ.do?uri=OJ:L:2008:354:0001:0006:en:pdf.

Regulation 1332/2008. Regulation (EC) on food enzymes and amending Council Directive 83/417/EEC, Council Regulation (EC) No 1493/1999, Directive 2000/13/EC, Council Directive 2001/112/EC and Regulation (EC) No 258/97. Official Journal of the European Union L 354: 7-15. Available at: http://eur-lex.europa.eu/LexUriServ/LexUriServ.do?uri=OJ:L:2008:354:0007:0015:en:pdf.

Regulation 1333/2008. Regulation (EC) on food additives. Official Journal of the European Union L 354: 16-33. Available at: http://eur-lex.europa.eu/LexUriServ/LexUriServ.do?uri=OJ:L:2008:354:0016:0033:en:pdf.

Regulation 1334/2008. Regulation (EC) on flavourings and certain food ingredients with flavouring properties for use in and on foods and amending Council Regulation (EEC) No 1601/91, Regulations (EC) No 2232/96 and (EC) No 110/2008 and Directive 2000/13/EC. Official Journal of the European Union L 354: 34-50. Available at: http://eur-lex.europa.eu/LexUriServ/LexUriServ.do?uri=OJ:L:2008:354:0034:0050:en:pdf.

Regulation 470/2009. Regulation (EC) of the European Parliament and of the Council of 6 May 2009 laying down Community procedures for the establishment of residue limits of pharmacologically active substances in foodstuffs of animal origin, repealing Council Regulation (EEC) No 2377/90 and amending Directive 2001/82/EC of the European Parliament and of the Council and Regulation (EC) No 726/2004 of the European Parliament and of the Council. Official Journal of the European Union L 152: 11. Available at: http://eur-lex.europa.eu/LexUriServ/LexUriServ.do?uri=OJ:L:2009:152:0011:0022:en:pdf.

RIVM, 2006. Risicobeoordeling inzake aanwezigheid van LLRice 601 in geïmporteerde rijst (in Dutch), 13 september 2006. Available at: http://www.vwa.nl/portal/page?_pageid=119,1639827&_dad=portal&_schema=portal&p_file_id=12446.

SCFCAH, 2005. Standing committee on the food chain and animal health, section on toxicological safety & section on general food law, summary record of the meeting of 14 February 2005. Available at: http://ec.europa.eu/food/committees/regulatory/scfcah/general_food/summary14_en.pdf.

SCFCAH, 2006. SANCO – D1(06)D/413447. Summary record of the standing committee on the food chain and animal health held in Brussels on 14 December 2006, section toxicological safety of the food chain. Available at: http://ec.europa.eu/food/committees/regulatory/scfcah/toxic/summary23_en.pdf.

Scientific Committee for Food, 1984. Report of the Scientific Committee for food on sweeteners (Opinion expressed in 1984), 16th Series, 1985. Available at: http://www.europa.eu.int/comm/food/fs/sc/scf/reports/scf_reports_16.pdf.

Scientific Committee for Food, 1997. Minutes of the 107th meeting of the Scientific Committee for food held on 12-13 June 1997 in Brussels. Available at: http://ec.europa.eu/food/fs/sc/oldcomm7/out13_en.html.

Spengler, M. and Sommerauer, B., 1989. Toleranz und akzeptans von isomaltulose (Palatinose®) im vergleich zu saccharose bei 12-wüchiger oraler gabe von aufsteigenden dosen (12-48 g) am gesunden probanden. Isomaltulose-Studie Nr. 101, Bayer Bericht Nr. 17792 (P) vom 7/3/1989. Cited in: Lina, B.A.R., Jonker, D. and Kozianowski, G., 2002. Isomaltulose (Palatinose®): A review of biological and toxilogical studies. Food and Chemical Toxicology 40: 1375-1381.

Sträter, P.J. and Irwin, W.E., 1991. Isomaltulose. In: L. O'Brien Nabors and R.C. Gelardi (eds.). Alternative sweeteners (2nd rev. expanded ed.) Marcel Dekker; New York, pp. 299-307. Cited in: Application for the approval of isomaltulose, page 5, Available at: http://www.food.gov.uk/multimedia/pdfs/isomaltulose.pdf.

Tacken, G.M.L., Banse, M., Batowska, A., Nesha Turi, K., Gardebroek, C., Wijnands, J.H.M. and Poppe, K.J., 2009. Competitiveness of the EU dairy industry - innovative and global players, but losing market share. LEI, The Hague, in press.

Takazoe, I. 1985. New trends on sweeteners in Japan. International Dental Journal 35: 58-65.

Tsuji, Y., Yamada, K., Hosoya, N. and Moriuchi, S., 1986. Digestion and absorption of sugars and sugar substitutes in rat small intestine. Journal of Nutrition Science and Vitaminology 32: 93-100.

Van der Meulen, B.M.J. and Van der Velde, M., 2005. Legislation and food innovation. In: W.M.F. Jongen and M.T.G. Meulenberg (eds.). Innovation in agri-food systems. Product quality and consumer acceptance. Wageningen Academic Publishers, Wageningen, the Netherlands.

Van der Meulen, B.M.J. and Van der Velde, M., 2008. European food law handbook, Wageningen Academic Publishers, Wageningen, the Netherlands.

Van der Meulen, B.M.J. and Freriks, A.A., 2007. Beastly bureaucracy. Animal traceability, identification and labelling in EU Law. Journal of Food Law & Policy issue 1/2006. Available at: http://www.law.wur.nl/NR/rdonlyres/1B763C73-1A3C-4902-B8A3-24D4A458C2AB/67616/MeulenvanderFreriks2006BeastlyBureaucracy_AnimalTr.pdf.

Van der Meulen, B.M.J., 2008. Judge Nelson Timothy Stephens lecture: the EU regulatory approach to GM foods. The Kansas Journal of Law & Public Policy 16: 286-323.

Van Horne., P.L.M., Oosterkamp, E.B., Hoste, R., Puister, L.F. and Backus, G.B.C., 2006. Herkomst-aanduiding van vlees: nationaal of Europees? (in Dutch) LEI-report 6.06.13, The Hague, the Netherlands. Available at: http://www.lei.wur.nl/NL/publicaties+en+producten/LEIpublicaties/?id=735.

WHO, 1987. Principles for the safety assessment of food additives and contaminants in food. World Health Organisation (WHO), International Programme on Chemical Safety (IPCS); Geneva, Switzerland. Environmental Health Criteria, 1987, No. 7.

Will, M. and Guenther, D., 2007. Food quality and safety standards, as required by EU Law and the private industry with special reference to the MEDA countries' exports of fresh and processed fruits & vegetables, herbs & spices. A practitioners' reference book, 2nd edition. GTZ, p. 16. Available at: http://www2.gtz.de/dokumente/bib/07-0800.pdf.

Wijnands, J.H.M., Van der Meulen, B.M.J. and Poppe, K.J., 2007. Competitiveness of the European food industry. An economic and legal assessment. Office for Official Publications of the European Communities. Available at: http://ec.europa.eu/enterprise/food/competitiveness_study.pdf.

Würsch, P., 1991. Metabolism and tolerance of sugarless sweeteners. In: A.J. Rugg-Gunn (ed.). Sugarless: the way forward. Elsevier Applied Science; New York, USA, pp. 32-51.

Yamaguchi, K., Yoshimura, S., Inada, H., Matsui, E., Ohtaki, T. and Ono, H., 1986. A 26-week oral toxicity study of palatinose in rats. Oyo Yakuri 31: 1015-1031.

Ziesenitz, S.C., 1986. Carbohydrasen aus jejunalmucosa des Menschen [Carbohydrases from the human jejunal mucosa]. Zeitung fur Ernahrungswissenschaft 25: 253-258. Cited in: Würsch, P., 1991. Metabolism and tolerance of sugarless sweeteners. In: Rugg-Gunn, A.J. (ed.). Sugarless: the way forward. Elsevier Applied Science; New York, USA, pp. 32-51.

Index

L

label – 87, 91
- private – 91, 92, 97

labelling – 15, 19, 20, 22, 45, 69, 71, 75, 81-83, 87, 92, 97, 100
- Country of Origin Labelling (COOL) – 78, 79, 83, 88-90, 92, 97
- EU proposal – 85
- law – 71, 73
- legislation – 20, 69, 72
- new proposal – 70
- new proposal, Article 35(2, 3 and 4) – 85
- nutrition – 70, 73
- proposal – 92
- provision – 73
- requirement – 69, 72, 81

lack of accessibility – 19

law
- European – 11, 23, 24, 41, 66, 71
- private – 35, 51, 75, 103
- product liability – 91, 92
- public – 35, 49, 75, 103, 105

legal certainty – 59
- lack of – 19, 57

legal texts
- consolidation – 33
- official binding – 33

legislation – 15, 17, 19, 20, 22, 25, 27, 29, 35, 37, 60, 67, 83, 95, 99
- European – 25, 32, 33
- Member State – 38
- national – 25, 72, 86
- procedure – 32
- system – 33

liability – 91, 103

M

maximum residue level (MRL) – 63
MRL – *See* maximum residue level

N

nangai nut – 53
new Novel Foods Regulation (NNFR) – 54-56
- Article 1(3) – 54

– Article 2(3) – 54
NFR – *See* Novel Foods Regulation
NNFR – *See* new Novel Foods Regulation
NOAEL – *See* no observed adverse effects level
no history for food use – 110, 115
no observed adverse effects level (NOAEL) – 59, 61
novel food – 19, 20, 37, 43-52, 73, 108, 109, 111-117
- catalogue – 48, 57
- ingredient – 111
- proposal – 41
- scheme – 48

Novel Foods Regulation (NFR) – 20, 22, 32, 37, 40, 45, 46, 48-50, 52-54, 108, 114, 116
- approval – 51
- Article 1 – 45
- Article 1(2) – 47, 109
- Article 1(2)(f) – 113
- Article 1(3) – 47
- Article 5 – 117
- case study – 20, 31, 34, 48, 50-52, 108
- procedure – 51

novel product – 53
novel protein – 116

O

Oca (*Oxalis tuberosa*) – 47
Official Journal – 41, 45
origin – 75, 79-83, 86
- certificate – 79
- designation – 86, 87
- European – 97
- geographical – 84
- indication – 86, 87
- labelling – 79, 83, 92, 97
- marking – 76, 78, 80, 81, 86, 87, 89
- Member State – 97
- of goods – 79
- of products – 73, 76, 77
- requirements (ORs) – 76, 90
- rules of - (ROOs) – 77, 78, 90

ORs – *See* origin requirements